Drawing by Wilbur Wright on brown wrapping paper shows top, side, and head-on views of the Wright brothers' first engine-driven flying machine. The triangle _____ *of the double vertical* _____ *piloted this cra* _____ *rth Carolin* _____ *, powered fli* _____

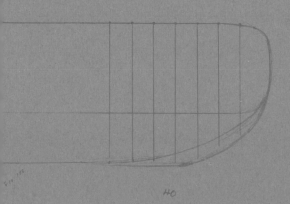

40

36

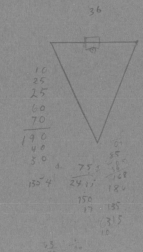

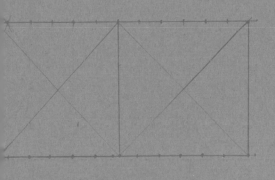

Foreword by Leonard Carmichael,
Vice President for Research and Exploration,
National Geographic Society, Washington, D. C.
Melvin M. Payne, President
Melville Bell Grosvenor, Editor-in-Chief
Gilbert M. Grosvenor, Editor

NATIONAL AIR AND SPACE MUSEUM, SMITHSONIAN INSTITUTION

A 1902 forerunner of the Wright brothers' powered craft, No. 3 glider lifts above dunes near Kitty Hawk, North Carolina, with

Those Inventive Americans

Produced by the National Geographic Special Publications Division,
Robert L. Breeden, Chief

Wilbur at the controls. After a thousand such flights, the brothers began experiments with engine-driven flying machines.

Cathedral Home School

THOSE INVENTIVE AMERICANS

Published by
The National Geographic Society

MELVIN M. PAYNE, *President*
MELVILLE BELL GROSVENOR, *Editor-in-Chief*
GILBERT M. GROSVENOR, *Editor*
ROBERT L. CONLY, *Consulting Editor*

Foreword by
LEONARD CARMICHAEL, *Vice President for Research and
Exploration*

Contributing Authors
MICHAEL AMRINE, EDWIN A. BATTISON, JOSEPH J. BINNS,
ROBERT V. BRUCE, PAUL DOUGLAS, ROBERT EVETT,
RONALD M. FISHER, JOHN GREENYA, MARY ANN HARRELL,
ROBERT W. HOLCOMB, HOWARD J. LEWIS, ARTHUR P.
MILLER, JR., CARROLL W. PURSELL, JR., TEE LOFTIN SNELL

Prepared by
The Special Publications Division

ROBERT L. BREEDEN, *Editor*
DONALD J. CRUMP, *Associate Editor*
MARY ANN HARRELL, PHILIP B. SILCOTT, *Manuscript Editors*
MARJORIE W. CLINE, MARGERY G. DUNN, ELOISE T. LEE,
MIRIAM D. PLOTNICOV, ELIZABETH C. WAGNER, PEGGY D.
WINSTON, *Research;* TUCKER L. ETHERINGTON, *Research
Assistant;* MARGERY G. DUNN, *Style*

Illustrations
WILLIAM L. ALLEN, *Picture Editor*
JOSEPH A. TANEY, *Art Director*
JOSEPHINE B. BOLT, *Assistant Art Director*
URSULA PERRIN, *Design Assistant*
WILLIAM L. ALLEN, RONALD M. FISHER, PHILIP KOPPER,
H. ROBERT MORRISON, GERALD S. SNYDER, *Picture
Legends*
ARTHUR LIDOV, STANLEY MELTZOFF, *Illustrators*

Production and Printing
ROBERT W. MESSER, *Production Manager*
ANN H. CROUCH, GAIL FARMER, *Production Assistants*
JAMES R. WHITNEY, JOHN R. METCALFE, *Engraving and
Printing*
SUZANNE J. JACOBSON, RAJA D. MURSHED, DONNA REY NAAME,
JOAN PERRY, SUZANNE B. THOMPSON, *Staff Assistants*
MARTHA K. HIGHTOWER, BRIT AABAKKEN PETERSON, *Index*

Copyright © 1971 National Geographic Society. All rights reserved.
Reproduction of the whole or any part of the contents without written
permission is prohibited.

Standard Book Number 87044-089-6
Library of Congress Catalog Card Number 75-125340

PAGE 1, COURTESY CHARLES H. TOWNES

DAVID BRILL

*Up to her neck in ingenuity, Betty Galloway of
Georgetown, South Carolina, beams at a bubble
produced by a toy she patented at age 10. Immersing
the device forces air against a soap film, thus forming
the bubble. Page 1: Stylized laser beams, symbo-
lizing man's inventive genius, shoot skyward in
the illustration for the 1964 Nobel Prize parchment
presented to American physicist Charles H. Townes.*

Foreword

IT SEEMS particularly appropriate that the National Geographic Society should publish *Those Inventive Americans,* for the work of these men has changed the character and face of our country. This book emphasizes how original some human beings are and how their creativeness makes life easier for the rest of us.

In these pages, we see that invention often depends on scientific research. When academic science registers an advance, applications may occur almost simultaneously to inventors in different countries. Sometimes it is impossible to determine who was first, because a number of workers hit upon the invention independently. In such cases, the answer was actually floating, as it were, in the intellectual climate of the day.

This complexity makes it difficult to compare the inventiveness of nations. Nevertheless the United States well deserves its reputation for Yankee ingenuity. The 1,800 patents issued weekly by the U. S. Patent Office attest this.

In this volume it is obviously impossible to consider more than a few inventors who have changed life not only in America, but also in the entire world. Some—Samuel F. B. Morse, Alexander Graham Bell, Thomas Edison—immediately come to mind. Others—John Fitch, Herman Hollerith, Vladimir Zworykin—are less generally known for their help in reshaping society.

We often forget how swift this change has been. As I looked at the engravings of the Philadelphia Centennial of 1876 in this volume, I thought of my father. As a high-school student, he helped his physics teacher install arc lights in Philadelphia for the Centennial. Now, arc lights for general illumination seem quaint indeed.

It is a mistake to believe that all inventors are motivated only by desire for financial gain. Joseph Henry, the first Secretary of the Smithsonian Institution, was a great scientist and inventor, but he refused to patent his discoveries. He wished them to be used for the general welfare of mankind. His aim was always to help struggling fellow inventors. His interest was only in the increase of knowledge and in the diffusion of the results of scientific investigation.

When we come to recent inventions, such as television and the transistor, we find that in almost all cases the invention is a result of many brains, working together or in competition, to reach a desired objective.

The research facilities of a modern industrial organization would have startled early inventors. For example, what would Clarissa Goodyear think of the Bell Telephone Laboratories? Her husband Charles used her small kitchen as his laboratory. But the desire for service that inspired Goodyear still activates workers today.

This should not suggest that the day of the lone inventor is over. In garages and basements across our country, men and women today are working as diligently as ever—often, now, in the cause of preventing pollution.

For one of the results of our inventiveness has been the harming of our natural environment. We are now becoming conscious of the urgent need to restore and preserve healthful living conditions. This is inspiring a new group of inventors. Engineers are trying to produce an automobile engine that operates without contaminating the air; chemists are laboring to produce a gasoline additive that will increase automotive efficiency without toxic by-products.

All this ferment of invention makes it seem that we may be at the threshold of a great new creative age. Our scientific and constructive skill has made it possible for men to navigate in space. This same capacity, we hope, will mean that society will thrive in a world more adapted to effective human living than at any time since the first good flint ax.

Invention is not limited to the great men dealt with in this book. Rather, the capacity for creating is shared in a measure by everyone. It may even be that the reading of this book will inspire the inventiveness that lives in all of us.

LEONARD CARMICHAEL

Contents

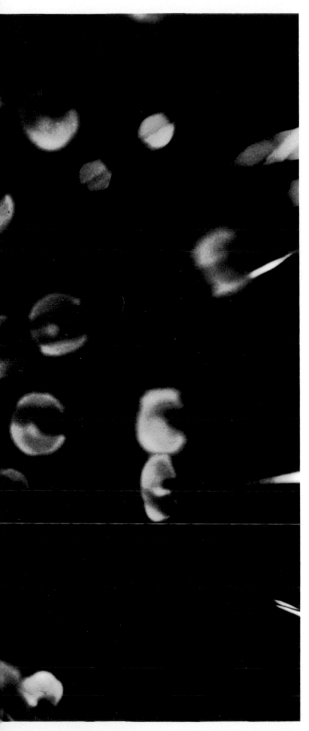

Copper wires spray from the end of a coaxial cable three inches in diameter that can carry 16,000 telephone conversations at once. Inside, plastic disks center the wires in copper tubes that shield signals from electrical interference. The name comes from the axis, or center, shared by the wires and tubes.

AMERICAN TELEPHONE AND TELEGRAPH CO.

Inventors of a New Nation: the Beginners

1

F IRST AND GREATEST of American inventions was, of course, the Republic itself. After July 1776 a new pattern of revolution took shape; other peoples tired of kings, other colonies eager for independence would fight and set out to govern themselves — with the American story to stir their own creative spirits.

Within two decades the successful revolutionaries of the United States would invent a dramatically new form of constitutional government and — with much struggle and tinkering — make it work.

Yet nothing in particular indicates that the American colonists considered themselves especially inventive. They took pride in being "freemen," and when they singled out "innovations" they spoke not of new contrivances but of new wrongs and tyrannies by ministers of the Crown. Yankee brag, with its boasts of

Benjamin Franklin amuses Philadelphia neighbors with "electrical fire" — static electricity. The glass globe, spun against a grounded buckskin pad, generates a positive charge. This flows through the knobbed collector and chain into the glass — a crude capacitor — and attracts a negative charge from the ground to the outside of the glass. The negative charge flows through the girl's body and sparks from her finger to the positively charged collector.

STANLEY MELTZOFF

Experiments and Observations (right), an illustrated volume of Franklin letters published in England, outlines his theory of electricity. In the portrait, he turns toward a gadget he connected to one of his inventions, the lightning rod. A thunderbolt conducted harmlessly to the ground causes a brass ball to strike a pair of bells. Outside, an electrical storm damages nearby buildings unprotected by the rods.

ingenious inventions by the score, came later.

How, in fact, could the rebels of '76 have supported claims to American inventiveness, if the idea had occurred to them?

Settlers with boundless wilderness to clear had developed a sharper, better-balanced ax. Shipbuilders along a tricky coast had developed the schooner. Gunsmiths in Pennsylvania had developed a notably accurate rifle from German prototypes. But such devices would prove more about response to frontier conditions than about technological advance.

Patent records—an indispensable if imperfect gauge of inventiveness—would offer little help. No colony granted many patents; some colonies granted none.

Fortunately, however, the first man to receive a patent for machinery in British North America admirably represents a much-neglected group, the skilled immigrants who kept American communities from lagging hopelessly far behind the new technologies of Europe.

He was an ironworker named Joseph Jenks; the settlers of Massachusetts Bay, needing a man to help them build a foundry and forge to produce basic tools and utensils, induced him to leave England and join them. In 1646 the General Court—the legislature—gave him a 14-year monopoly on watermills for "speedy dispatch of much worke with few hands."

Much work with few hands ... that lack of labor would mark the American economy for generations. The desire for "speedy dispatch" would spur inventors on and become, eventually, a national characteristic.

Thus wider fields than those of England and the perennial shortage of workmen inspired Jenks's invention of 1655, a brilliantly improved scythe "for the more speedy cutting of grasse." He replaced the short, thick English blade with a long and thin one, strengthened with a bar of iron; and his improvement became the new standard model.

Not every immigrant craftsman would turn out

to be an inventor as well, but for decades America would depend on men like Jenks: immigrant artisans, mechanics, engineers.

Such immigrants, as historian Brooke Hindle points out, should stand beside the traditional heroes of storybooks and textbooks, "American whittling boys inventing their way ... from the colonial world of hand production to the new world of machine production."

In a typical protest on the eve of revolt, Virginia patriots would resolve not to buy British imports —and prove the backwardness of home industry by their list of necessary exceptions: cloth, medicine, gunpowder, lead, tools "for the handicraft arts and manufactures."

Odd as it seems today, America was too rich for industry, as Benjamin Franklin had observed in 1755. An able man could acquire land easily and support his family, and never be "poor enough to ... work for a master" at comparatively low wages.

Franklin remains the only immortal among American inventors from that colonial and pre-industrial age. For all his unequaled sense of the practical, and the versatility that led one admirer to say he was "almost like a one-man Renaissance," nevertheless he invented nothing that could be called machinery but a clock.

In 1774, when the King's solicitor-general attacked him as "inventor and first planner" of Britain's trouble with her Massachusetts colony, Franklin enjoyed world renown as an "electrician"—in those days, a scientist working in a fascinating, if confusing, realm of no great practical use. "Electrical fire" or "fluid," at the time Franklin began to investigate it in the 1740's, meant the sparks from rubbed glass globes or tubes: an artificial curiosity that an experimenter might use to entertain persons of rank. With two cork balls suspended on silk threads, he could show that they would draw together if one ball received a negative charge and the other a positive charge—or swing away from each other if both received the same charge. All the electri-

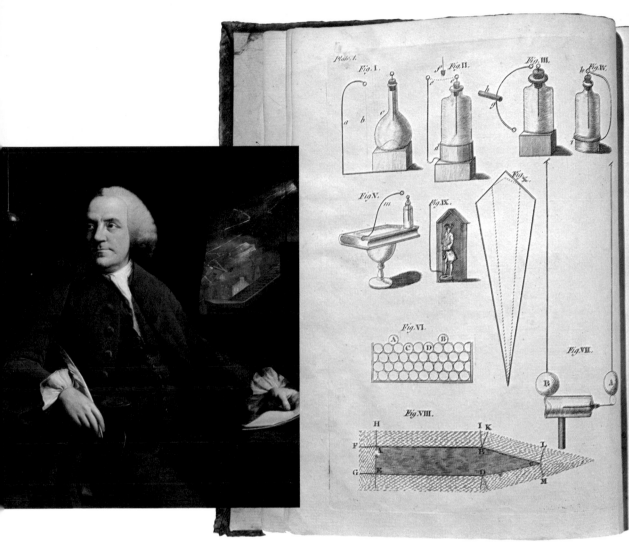

THE FRANKLIN INSTITUTE, PHILADELPHIA (RIGHT); MASON CHAMBERLIN, 1762, PHILADELPHIA MUSEUM OF ART

cians since have worked with the phenomena he demonstrated.

Franklin gave the subject much of its English vocabulary, an explanation of the new Leyden jar (the first condenser, or capacitor, that stored an electric charge), and a coherent theory.

Above all, by proving electricity identical with lightning, he gave it universal importance as a basic fact of nature. "Thereafter," writes physicist-historian I. Bernard Cohen, "every experimenter rubbing glass tubes in his laboratory knew that he was studying cosmic forces on a small scale."

And the honest glamor of Franklin's discoveries concerning lightning flickered around his plain fur hat when he landed in France in December 1776 as chief diplomat of the Revolution.

A much more dubious romance surrounds the colonies' other notable physicist-inventor — if America can fairly claim Benjamin Thompson of Massachusetts.

Born in Woburn in 1753, he narrowly escaped electrocution trying to repeat Franklin's kite experiment in 1771, narrowly escaped a tar-and-feathering by a mob of Concord patriots in 1774, and, it seems, narrowly escaped exposure in political intrigue all his life.

During the Revolution he chose to serve the King in name, his powerful patron Lord George Germain in fact, and — as always — himself. When

11

American spinning wheel and British Boulton and Watt steam engine symbolize the economies of the two nations in the late 18th century. Gears and a flywheel (right) translate the engine's piston strokes into rotary motion to drive machinery for grinding

ARTHUR LIDOV

flour, polishing metal parts—or spinning yarn. Thus, while Europeans developed machines, Americans whittled frontier implements. England encouraged the technology gap; she considered her colonies mainly sources of raw materials and markets for finished products. Abundant land also contributed to America's industrial backwardness—for who would labor for a master when he could farm his own property? But this very spirit of independence soon fostered the growth of American inventiveness.

Inventor and opportunist, American-born Benjamin Thompson won fame in Europe. Napoleon, a brilliant artillerist, once refused the scientist, a master of ballistics, entrance to France because of his work for England. But Thompson later enjoyed the Emperor's friendship and a Paris estate, even as French troops (right) fought the British in Spain. Between intrigues he investigated the properties of heat and light.

Benjamin Thompson (1753-1814)
Physicist, engineer, mercenary

fellow Americans in the past. But at least the count and the Emperor could discuss artillery with the enthusiasm of two experts.

In long experiments on gunpowder and on cannon-boring, Thompson had studied the properties of heat and made valuable contributions to science. Eager to turn his discoveries to practical good, he virtually invented the modern kitchen.

Preparing food at an open fire, as he said, would "cook the cook more than the food." He wrapped the fire in an insulated box, thus inventing the kitchen range; later he invented the double boiler as well. To provide good coffee as a substitute for strong drink, he devised the first drip coffeepot.

Central heating, a lost art since Roman times, returned to everyday life when he designed a system of steam radiators.

Despite these and many other solid achievements, Benjamin Thompson — Count Rumford — remains as obscure as any successful inventor in history. Proud of her rebel patriots, scornful of loyalists, America has naturally neglected him. His dream of returning to his native land to establish a national military academy foundered when the officials of the United States recalled his unsavory role in the War of Independence: a typical episode in his bizarre, adventurous life.

And the count — as haughty as Franklin was genial, as difficult as Franklin was easy — exasperated almost everyone he met. When he died in 1814, says his biographer, physicist Sanborn C. Brown, those who knew him "just forgot him as fast as possible."

Either politics or personality can make an inventor's fame dwindle to the vanishing point.

But Thompson's career sheds its own light on the country he decided to abandon. Here he could not have found technicians or riches or the patronage of powerful princes to support his work.

In his time America was — though nobody used the term — an underdeveloped country.

a French spy was caught with secret plans of the British fleet in 1781, gossip named Thompson as his informant. Thompson promptly left London for America to recruit a loyalist unit called the King's American Dragoons.

After the war he found it prudent to serve Karl Theodor, Elector of Bavaria, a prince who could appreciate the worth of a scientist-and-soldier-of-fortune and ennoble him as Count Rumford of the Holy Roman Empire.

In 1804 the restless count, having made enemies elsewhere, thought it best to settle in Paris. There Napoleon, Emperor of the French, made much of him — but always with a certain reserve, as if he believed the rumors that his friend was a British spy. An aura of espionage clung to the man who had indeed spied on his

"FRENCH ARMY CROSSES THE SIERRA DE GUADARRAMA," NICHOLAS-ANTOINE TAUNAY, CA. 1812, MUSÉE DE VERSAILLES, EKTACHROME DES MUSÉES NATIONAUX (ABOVE); THOMAS GAINSBOROUGH, 1783, FOGG ART MUSEUM, HARVARD UNIVERSITY, BEQUEST OF EDMUND C. CONVERSE

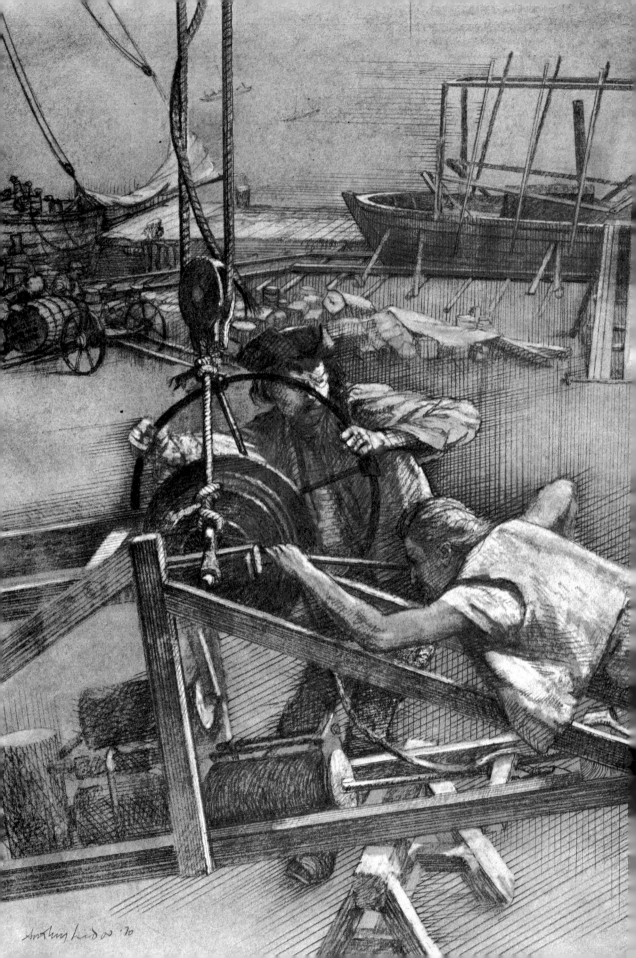

Frontier of Power: Steam and Its Heroes

2

IN APRIL 1776, patriots in 13 United Colonies were coming to accept the radical idea of independence; the forces of Gen. George Washington were guarding the strategic city of New York; and one of his soldiers, a young Harvard graduate named Isaac Bangs, was recording the strange sights he saw there.

He found a "mystery" at the new waterworks reservoir east of Broadway: how the machine that ran the pump was "actuated and kept in motion." As he "at length discovered, with surprise," it worked "by the power of boiling water."

The astonished Bangs had seen his first steam engine—one of two known to exist in the whole country then—and his careful description of its "strong copper tube" (the cylinder) and "moveable stopper" (the piston) reveals the type: Thomas Newcomen's.

Newcomen, a Devonshire ironmonger, had

Inventor John Fitch strains to adjust the axletree on his steam engine in Philadelphia in May 1787. His partner, German-born watchmaker Henry Voight, holds the flywheel used to smooth the jerky motion of the engine. Fitch built the seven-ton engine for the 45-foot skiff in the background, hoping to have the first successful steamboat. This version of the engine failed—as would Fitch's dreams of glory.

ARTHUR LIDOV

James Watt (lower right) conceives a more efficient steam engine in 1765 at Glasgow University in Scotland while repairing a model of one invented by Englishman Thomas Newcomen. Newcomen's engine alternately heated and cooled its own cylinder as it produced, then condensed, steam. To eliminate the waste in heating the cylinder after each stroke, Watt used a separate condenser. In 1801 the waterworks engine house in Philadelphia's Centre Square (right) held an American engine patterned after Watt's.

installed his first steam engine at a coal mine near Birmingham, England, in 1712, to supply power for a lift pump to raise water from the shaft. A neglected figure by comparison with his celebrated successor James Watt, he had produced, as a modern scholar says, "one of the great original synthetic inventions of all time."

Ponderous and feeble by present standards, the Newcomen engine was a marvel of power and reliability next to those medieval stalwarts, the waterwheel and the windmill, and the ancient resources of animal or human muscle.

A boiler provided steam that rose to fill and heat a cylinder. There injections of cold water condensed the steam, creating a partial vacuum. Then the weight of the atmosphere pushed the piston into the cylinder, supplying energy for a pumping stroke that could raise water 90 feet.

For 60 years this "atmospheric engine" reigned unchallenged at mines or waterworks in England, Hungary, Germany, France, and Sweden. In British North America, however, only one copper mine—in New Jersey—used such an engine, installed in 1755, and not until 1774 did New York City, with 22,000 residents, decide it needed the machine seen by Isaac Bangs.

Alert young men like John Adams and Thomas Jefferson studied reports of the "fire engine" as early as the 1760's, but steam power remained hearsay to most Americans long afterward.

The Revolutionary War shut them off from knowledge of Watt's contributions to steam engineering. Watt—who learned his art on a Newcomen model—offered his first engines for sale in 1776. To avoid the alternate heating and chilling of a cylinder, he had added a separate vessel where the steam was condensed. The cylinder stayed hot, the condenser cool. This saved half the steam produced by the boiler and doubled the thermal efficiency of the best Newcomen engines. By the time the war ended, Watt and his partner Matthew Boulton were developing even better engines, but in 1785 Parliament made all export sales illegal.

Thus the American pioneers of steam, struggling to adapt this new force to the service of their new republic, worked—as many confessed —in ignorance. Usually the local artisans they hired had never even seen an engine. They needed resources of mechanical skill and industrial capacity and investment capital that simply did not exist in their time.

Under such conditions Oliver Evans from New Castle County, Delaware, earned his proud epithet "the American Watt," inventing engines that used steam itself to push the piston at pressures three to five times that of the atmosphere.

Rating the performance of such early engines gave the builders almost as much trouble as designing or producing them. But probably Evans knew the standard that Watt had devised—the modern horsepower unit, the ability to raise 33,000 pounds one foot in one minute—when he announced in 1812 that his engine of 24 horses could grind 240 bushels of grain or saw 5,000 feet of boards in a 12-hour day.

The future belonged to high-pressure engines of this type. With their help the American economy changed almost beyond recognition within half a century.

By 1838 the United States could report at least 3,010 steam engines in operation, scattered west to the Mississippi, some 800 in boats, about 350 in locomotives, the rest of them stationary: in textile and paper mills, metalworks, sawmills, gristmills, printing plants, breweries, saltworks, rice mills in Georgia, sugar mills in Louisiana, cotton gins throughout the South. And most of these engines were American made.

At least one of the far-sighted men of the 1790's lived to see the steamboat successful: Samuel Morey of Orford, New Hampshire. Finding it impossible to get backing for his sidewheeler of 1793, he had sensibly turned to other projects. He came to regret his decision and denounced the Johnny-come-latelies in terms that innumerable inventors have echoed: "Damn their stomachs. Those cusses stole my invention."

PENNSYLVANIA ACADEMY OF THE FINE ARTS, PHILADELPHIA (ABOVE); COURTESY MRS. DOROTHY HENRY, INSTITUTION OF BRITISH ENGINEERS, LONDON

Twelve paddles digging steadily at the water, a steamboat built by Connecticut-born John Fitch chugs past Philadelphia's Independence Hall in August 1787. Cranks pull the paddles through the water, lift them out, and carry them forward for another stroke as spectators watch the trial run. The following year Fitch and his partner tried paddles at the stern, as in their French patent application below. The boat worked, but it failed financially.

THE CONTINENTAL INSURANCE COMPANY'S COLLECTION (ABOVE); SMITHSONIAN INSTITUTION

JOHN FITCH and JAMES RUMSEY
By Paul Douglas

ARTHUR LIDOV (BELOW); WEST VIRGINIA DEPARTMENT OF ARCHIVES AND HISTORY

John Fitch (1743-1798), James Rumsey (1743-1792)
Rivals in the race for the first steamboat

T O THE PEOPLE of Bardstown, Kentucky, in the year 1798, the tall, gaunt figure of John Fitch—dressed in threadbare old clothes and muttering about steamboats, Benjamin Franklin, and "the ignorant boys of Congress"— was a strange figure indeed. He had arrived in the town two years before, a penniless and dejected old man, and persuaded an innkeeper to give him room, board, and a pint of whiskey a day in return for 150 acres of land which Fitch claimed would soon be granted him by the courts.

Few, perhaps none, in that Kentucky town knew they were witnessing the last days of one of the great American inventors. In 1790, a full 17 years before the more famous Robert Fulton's achievements on the Hudson, John Fitch built a steamboat that regularly plied the waters of the Delaware between Philadelphia and Trenton, New Jersey. He called himself "little Johnny Fitch," this brilliant yet pathetic man who sacrificed all he had in pursuit of the steamboat.

Born in 1743, near Windsor, Connecticut, John was the fifth of a farmer's six children. When he was five years old, John badly burned his hands putting out a fire in the kitchen. His only reward was punishment from his brother Augustus, who must have thought John had started it. "This," he wrote forty years later, "... being what I may call the first act of my life, seemed to forebode the future rewards I was to receive for my labors through life...."

Apprenticed at 18 as a clockmaker to the Cheney brothers of East Hartford, he was not properly taught his trade. This injustice caused him to break his apprenticeship, and more than ever to distrust authority.

In 1769, after 13 months of a quarrelsome marriage, he gloomily abandoned his pregnant wife Lucy and their infant son. Fitch later remarked, "I know nothing so perplexing and vexatious to a man of feelings as a turbulent Wife and Steamboat building.... [F]or one man to be teised with Both, must be looked on as the most unfortunate man of the world."

Fitch then worked his way down to New Jersey, cleaning clocks as he went, and in Trenton made brass buttons, repaired watches, and learned the craft of silversmithing. By the outbreak of the Revolutionary War Fitch was a well-known silversmith—but the war would see to it that his success was short-lived.

The British wrecked his workshop; when his fellow patriots refused him a commission in the American service, he turned to surveying and buying land in Kentucky. In March 1782, on a surveying expedition near present-day Marietta, Ohio, Fitch and nine other men were captured by Indians. After a strenuous journey, the captives were turned over to the British and held at a prison depot on the St. Lawrence River.

During this confinement Fitch felt his greatest sense of success. He obtained wood by hiring prisoners with buttons he made, and built "9 wooden timepeaces." Here was John Fitch, a prisoner in the wilderness, yet a successful craftsman! "In short," he said, "in about four months I had got to be as rich as Roberson Cruso." Fitch might have died a man satisfied

with his achievements if he could have lived out his days as an aristocrat among prisoners. In this tiny community he was recognized as a man of talent. It was only for a moment.

After his release from prison, Fitch, now penniless, went to Bucks County, Pennsylvania, and found work as a surveyor in the wild area from the Appalachian Mountains to the Mississippi. Different states had conflicting claims in this region and titles to land were clouded, but Fitch did his best to secure good acreage for himself and his associates. And his "Map of the Northwest Parts of the United States of America" had the clear distinction of being the only map made, engraved, and printed (on a neighbor's cider press) by the same man.

On a clear and warm Sunday in April 1785, Fitch and a friend were walking home from church near Neshaminy in Bucks County, when a pang of rheumatism struck him "pretty severely" in the knee just as a horse-drawn carriage passed by. At this moment Fitch, who had rented out his own horse, got his idea for a steam engine to propel a vehicle: "What a noble thing it would be if I could have such a carriage without the expense of keeping a horse!"

For a week he struggled over the plans for a steam carriage, but abandoned them when he realized the difficulties posed by the primitive roads of the day. He then turned to the steamboat, his vision of river navigation that was to consume the rest of his life—a vision that would lead to poverty, despair, and suicide.

After showing his drawing for a steam engine to a friend, he was "considerably chagrianed" to learn that others had thought of the steam engine before he had. But he persevered, and in attempts to finance the construction of a boat Fitch unsuccessfully petitioned prominent men of Pennsylvania, Congress, and the Minister of Spain, who then lived in New York.

In September 1785 Fitch presented a drawing and a model of his boat, which was propelled by a series of paddles on an endless chain, to the American Philosophical Society at Philadelphia. He also appealed to Benjamin Franklin, the Society's most famous member, for his endorsement. When his letter was not answered, Fitch must have seen in the statesman another figure of authority, another Cheney trying to cheat him, for he accused Franklin of "being desirous of gaining the Honour to himself."

When Fitch finally met Franklin in Philadelphia, the stately old man was surprised by the volatile creature who stood before him and in-

ARTHUR LIDOV

Enraged and ill-dressed, John Fitch berates Virginian James Rumsey on a Philadelphia street in 1788, accusing him of trying to steal credit for the steamboat. After Rumsey published his claim for the inven- *tion, Fitch promptly confronted him with a paper of his own. In the "pamphlet war" that followed, each stoutly supported his own claims, but Robert Fulton would gain the enduring glory of recognition.*

sisted that his steamboat was a valuable invention. Fitch's problems with boilers, cranks, and paddles were minor compared with those of his abrasive personality, for he alienated all the men he should have impressed. Franklin, either to pacify or get rid of Fitch, drew out five or six dollars which he handed to his talkative and almost violent visitor. But Fitch felt that he had been slighted, and commented bitterly, "I esteem it one of the most imprudent acts of my life, that I had not treated the insult with the indignity

which he merited, and stomped the poltry Ore under my feet." If the most famous American inventor did not realize the worth of his steamboat, Fitch would seek others—and he did raise $300 by selling $20 shares.

Failing in his attempt to purchase a steam engine from England, Fitch was fortunate to meet a watchmaker in Philadelphia who had the mechanical knowledge that he lacked. Henry Voight, as Fitch described him, was a "Plain Dutchman, who fears no man, and will always speak his sentiments. . . ." To Fitch, Voight was no threat, no figure of authority like Franklin, but a plain man like himself who would show the world his worth and who would do it without manners and polite conversation. What did manners have to do with steamboats anyway?

Their first boat, worked by hand since the engine was not finished, proved a failure. As Fitch and Voight strained to turn the awkward combination of gears and cranks, the paddles flailed, the boat wallowed clumsily, and the Delaware River boatmen "scoffed and sneered."

Fitch bought a bottle of rum and drank himself to sleep in his room. What had he to show for his 43 years but a half-completed steam engine and the reputation of a fool?

Yet Fitch's spirits rose again as he gained exclusive navigation rights on several rivers and partial success with the engine installed in his 34-foot skiff. It worked, but—sadly—the little boat managed only three or four miles per hour against the current.

His despair growing as his subscribers dwindled, Fitch began to think that perhaps only future generations would ever recognize his talents. Throughout his difficult life, John Fitch was to feel that he was never properly rewarded for his invention, and he was right. His contribution to the development of the steamboat was recognized only long after he died.

When he tried again with a better engine in his first full-size boat, the 45-foot *Perseverance,* the Constitutional Convention was meeting in Philadelphia and a successful trial might attract influential men. Fitch could not converse politely in drawing rooms, but on the Delaware he might be convincing. Although the machinery weighed 7 tons, the 12 paddles moved the boat at 4 miles per hour on its second trial, and Fitch was certain that with an 18-inch cylinder instead of a 12-inch, his boat could reach 10 miles an hour.

Now, with success in sight, Fitch heard a disturbing rumor of one James Rumsey, who claimed *he* had invented the steamboat.

Rumsey's interest in a mechanically propelled boat began in 1784 at Bath, Virginia (now Berkeley Springs, West Virginia), where he was running a boarding house and building a number of houses for George Washington. He planned to propel his boat by manually operated "setting" poles placed on the riverbed and assisted by the force of the current. When General Washington stopped in Bath to examine his property in September 1784, he noted in his diary that he "was showed the Model of a boat . . . for ascending rapid currents by mechanism."

Rumsey's work on this boat was interrupted in 1785 when he accepted the position of Superintendent for the Patowmack Company, organized to bypass the falls and rapids on the Potomac River and open it to navigation. Despite Rumsey's being the "most skilled mechanician in the Two States," the canal failed financially.

Rumsey returned to his secret experiments with the steamboat. His engine, he hoped, would draw water into a tube on the bottom of the hull and force it out at the stern. Inhabitants of Shepherdstown, Virginia, lined the banks of the Potomac on a December day in 1787 to watch this mysterious boat prove whether its builder was "crazy Rumsey" or a man of genius. "I was standing next to General Gates," said one observer. "When she moved out and he saw her going off up the river against the current by the force of steam alone, he took off his hat and exclaimed, 'My God, she moves!' " The cheers of the crowd drowned out the noise of the engine.

And so by 1788 both Fitch and Rumsey had demonstrated that a steamboat could move under its own power. Although Rumsey's method of hydraulic propulsion was endorsed by Franklin and simpler than Fitch's awkward paddles, it was not so efficient. With each man confident that he should be called the inventor of the steamboat, a "pamphlet war" broke out between them.

Rumsey wrote that he was working on a steamboat as early as the fall of 1785. Fitch challenged

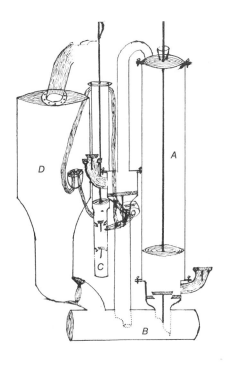

Drawing of a steam cylinder (A), condenser (B), air pump (C), and water tank (D) survives in one of six large notebooks compiled by John Fitch. This engine, in April 1790, propelled his boat at 8 miles an hour. Below, a Watt invention, the centrifugal or flyball governor, makes a steam engine self-regulating. Spun fast by excess speed, the whirling weights lift a valve, letting off steam and slowing the engine.

HISTORICAL SOCIETY OF PENNSYLVANIA (UPPER);
WILLIAM L. ALLEN, N.G.S. STAFF

Rumsey's date, saying "in the article of CONDENSATION I freely acknowledge he is my superior, having acquired the art of *condensing* (with the dash of his pen) one *whole year* into the compass of *six days.*"

Quite different from the abrasive and truculent Fitch, who must have looked and sounded like one of his steamboats clanking and chugging up the Delaware, Rumsey was cool-headed and realistic—he persuaded wealthy and influential men, including Dr. Franklin, to form a company called the Rumsean Society.

To Fitch's great chagrin, Rumsey was given $1,000 by this Society to study and forward his ideas in Europe and to build a boat in England with an engine from Boulton and Watt, the most experienced builders in the world. Whereas Fitch's fortunes fluctuated as wildly as the paddles on his steamboat, Rumsey's advanced with the regularity of his jet-propelled craft.

A country boy, Rumsey was amused by the formality he encountered in France. To his brother-in-law, he described a French social call: "I was obliged to be dressed in a black Coat-wescoat, breeches & Stockings, my hair handsomely dressed and powdered, and the hind part in a large black bag; by my Side a Sword; my hat in my hand; and . . . a lusty French Servant. . . . These are things . . . that at first, I had no idea was a necessary Conection of a Steam boat."

By the end of 1792 Rumsey had found, lost, and regained financial backing in England for the nearly completed *Columbian Maid* he was having built there. However, while attending a meeting to explain one of his inventions, Rumsey complained of a violent pain in his temples. The next evening, December 18, 1792, 49-year-old James Rumsey died. He never learned that the *Columbian Maid* achieved only a disappointing 4 miles per hour. If he had he might have abandoned his scheme for hydraulic propulsion and turned to more efficient methods.

With Rumsey in Europe, Fitch renewed his efforts to build a successful boat, and in July

National Park Service mule skinner walks beside one of a pair of mules that haul a sightseeing barge along the Chesapeake and Ohio Canal in Washington, D. C. In 1785 George Washington asked James Rumsey to construct a waterway paralleling the rapids-blocked Potomac River. An economic failure, the project had a successor—the present canal. Steamboats, a dream of Rumsey's, proved inferior to mules in pulling loads. Below, unused lock machinery becomes play equipment for a young visitor.

JOSEPHINE B. BOLT (ABOVE) AND H. ROBERT MORRISON, BOTH N.G.S. STAFF

1788, the *Thornton,* propelled by paddles placed at the stern, made a 20-mile trip from Philadelphia to Burlington, New Jersey—the longest distance any steamboat had yet traveled.

The spring of 1790 brought warm air and hope to the inventor, now ready with an improved boat. On April 16 the engine started, built up steam, and as Fitch exclaimed, "we reigned Lord High Admirals of the Delaware, and no boat in the River could hold its way with us, but all fell astern...." The *Thornton* moved at 8 miles per hour and Fitch was ecstatic: "Thus has been effected, by little Johnny Fitch and Harry Voight, one of the greatest and most useful arts that has ever been introduced into the world; and although the world and my country does not thank me for it, yet it gives me heartful satisfaction."

Fitch could not help the few words of self-pity. It had become a habit.

By autumn Fitch's steamboat had traveled almost 3,000 miles and carried 1,000 passengers. In spite of this unprecedented achievement,

which gives Fitch his fame today, the boat carried only about seven passengers per trip and the company lost money. A stagecoach could reach Burlington from Wilmington almost an hour and a half faster than the more comfortable steamboat, and if Americans couldn't have both comfort and speed, well, they would rather get there in good time than in good spirits.

This economic failure and the destruction of a new *Perseverance* in a storm shattered Fitch's hopes. A Philadelphian remembered the lonely and dejected old man of 49 years as "a troubled spectre, with down cast eye and lowering countenance, his coarse, soiled linen peeping through the elbows of a tattered garment."

After a few more futile attempts to interest people in his invention, Fitch made his way to Bardstown, where he hired a lawyer to press his land claims. In June 1798 poor John Fitch killed himself with an overdose of opium pills. Since the world refused to reward his great talents, he chose to leave the world.

Oliver Evans (1755-1819)
Inventor of the high-pressure steam engine

By John Greenya

ON A PLEASANT JULY DAY in Philadelphia in 1805, the world came to an end. Or so it must have seemed to any unlucky citizens who had not heard about the miracle scheduled to take place near the new waterworks in Centre Square. Even those who were forewarned found themselves startled or frightened by the spectacle, but the unsuspecting must have remembered how the great Puritan divine Jonathan Edwards had foretold the fate of sinners in the hands of an angry God: eternal fire.

For suddenly the summer air was rent by the sound of rumblings—regular, low-key explosions that issued from a workyard at Ninth and Market Streets, a few blocks from Centre Square. When the gates of the yard swung open, the noise increased and became an incredibly loud hiss. In the street all foot traffic stopped, at least momentarily, and dogs sought the haven of convenient alleys, their masters and mistresses not far behind.

The bold few who stood their ground saw, emerging from the gateway, a squared-off boat, in fact a scow, but one that was—could it be? —raised from the ground by creaking wheels set in a wooden frame. And it was coming into the street under its own power.

The workmen stood back and smiled proudly with accomplishment as the scow steamed past them, smoke billowing from its funnel and sparks streaming from its boiler.

Children hooted and scampered down Market Street behind the lumbering 30-foot-long beast, while mothers and maidservants hugged sobbing infants. Some people cheered from a distance while the boat that looked like a wagon clamored past at about the pace of a walking horse—though the comparison couldn't have been tested, for all the horses in the vicinity had bolted.

Lurching and chugging over the cobblestones, the machine crunched toward the square, as the little steam engine in its hull coughed with impressive regularity. People lined the way, and the braver ones got so close they could smell the animal fat packed around the axles as it heated with friction.

In front of the engine, wearing a wide smile across his well-fed face, sat a man of some fifty years; he lifted his beaverskin hat and doffed it to the crowd. Obviously it was one of his greatest moments, and not a few spectators said something like, "That Mr. Evans, perhaps he isn't so odd after all."

So one may reconstruct the ill-recorded debut of the country's first steam vehicle to travel on a city street. Its inventor, Oliver Evans, had invited the public to see it trundling around Centre Square. Probably, although the facts were lost, a simple gear train connected the engine and the road wheels, with a pilot wheel and tiller or ropes for steering.

After a few days on parade, it chugged down to the Schuylkill and settled ungently in the shallow water. Not long after, the tide swelled in to lift the weird craft from its wheeled framework and the ugly duckling, propelled by a paddle-wheel astern, steamed downriver.

A contemporary account explained its name: "Orukter Amphibolos," or Amphibious Digger; its purpose: "deepening the docks of the city of Philadelphia"; and its arrangement: "a steam

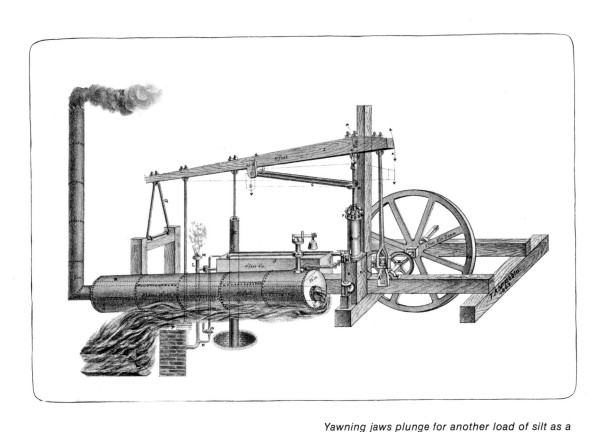

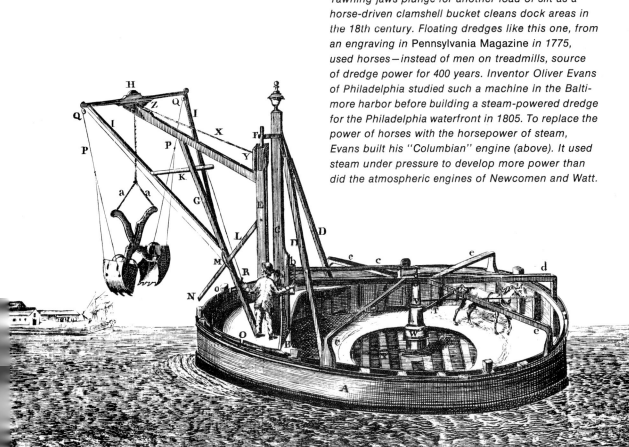

Yawning jaws plunge for another load of silt as a horse-driven clamshell bucket cleans dock areas in the 18th century. Floating dredges like this one, from an engraving in Pennsylvania Magazine *in 1775, used horses—instead of men on treadmills, source of dredge power for 400 years. Inventor Oliver Evans of Philadelphia studied such a machine in the Baltimore harbor before building a steam-powered dredge for the Philadelphia waterfront in 1805. To replace the power of horses with the horsepower of steam, Evans built his "Columbian" engine (above). It used steam under pressure to develop more power than did the atmospheric engines of Newcomen and Watt.*

FROM GREVILLE AND DOROTHY BATHE, "OLIVER EVANS," 1935, HISTORICAL SOCIETY OF PENNSYLVANIA (OPPOSITE); LIBRARY OF CONGRESS

STANLEY MELTZOFF

Triumphant Oliver Evans beams as his amphibious dredge, first steam-driven vehicle to travel on American

streets, wallows in shallows of the Schuylkill River after lumbering through Philadelphia in July of 1805.

engine, on board of a flat bottomed boat, to work a chain of hooks to break up the ground with buckets to raise it above the water, and deposite it in another boat to be carried off."

The Orukter Amphibolos did manage to fulfill Evans's contract to build a dredge for the city, which used it for three summers. But no records survive to indicate how much work it did or precisely why it was scrapped in 1809. To some degree, at least, it had managed to fulfill its inventor's lifelong dream, to see a steam engine propel a vehicle on land.

Nonetheless, it was not for this great dredgemobile that America and the world would—or should—remember Oliver Evans. His fertile imagination produced many inventions; the greatest was not the ponderous Orukter but rather its tiny heart, the little high-pressure steam engine that beat so steadily as the 20-ton craft waddled along.

In England, just the year before, Richard Trevithick, a Cornish mining engineer, had fired up his small locomotive and "pulled 10 tons of bar iron, 70 thrill-seekers and five cars along nine miles of Welsh colliery track in four hours." Like Evans, he had knowingly risked disaster in designing a high-pressure engine. Although the idea of such a system was far from new, the technique of building a safe high-pressure boiler still demanded more than most metalsmiths could offer at the time.

James Watt's cylinders held steam at a pressure of about 15 psi (pounds per square inch). Trevithick and Evans worked with pressures less than 50 psi. (Today's engines reach 3,000 psi, and beyond.) And Evans worked quite independently, building engines of his own original design.

Born in 1755 to a successful Delaware farming couple, Oliver Evans lived at a time when few Americans could understand the possibilities of what he and his peers called "strong steam." He met many of the men who made the Steam Era a part of the country's history. He was acquainted with both James Rumsey and "poor little Johnny Fitch"; he corresponded with and sometimes aided John Stevens, Robert Fulton, and Presidents Washington and Jefferson.

Because he felt others were infringing on his patents, he probably initiated more lawsuits than any other inventor of his day.

And, finally, Oliver Evans was able to see just what—to use his own words—"the celestial power of steam" promised for the future of the young American Nation.

Evans represented almost perfectly the dilemma of the American inventor in the late 18th and early 19th centuries: how to improve life for others while making a living for oneself. He was luckier than Fitch or Rumsey but not so well-born as Stevens nor so politic as Fulton, and thus time and time again he found himself both loving and cursing his gift.

For his day Oliver had a good education, and was apprenticed by his father to a wheelwright, which suited young Evans. He had always been attracted to mathematics and mechanics, and in this period the wheelwright and the millwright were the "engineers." During his apprenticeship he experimented in small ways, but a chance event put him onto his life's work.

In the Christmas season of 1772, when Oliver was 17, one of his brothers—he was near the middle of a line of 12 children—mentioned a trick the blacksmith's son had performed one afternoon. Taking a gunbarrel with the touchhole blocked, the lad poured in water and then wadded it tight. He put it into the smith's fire, and after the water had turned to steam the wad shot out with great force.

The others were impressed, but it immediately occurred to Oliver, as he wrote later, that here was "a power capable of propelling any waggon, provided that I could apply it."

This insight alone was an accomplishment. Typically, Evans tried to apply it, and "at length, a book fell into my hands, describing the old atmospheric engine"—probably Newcomen's. He read with astonishment that it used "the mere

N.G.S. PHOTOGRAPHER WINFIELD PARKS (ABOVE); LIBRARY OF CONGRESS

Morning fog rises around George Washington's gristmill near his Virginia home, Mount Vernon. Here Washington installed machinery designed by Oliver Evans. Below, an annotated engraving from Evans's book, The Young Mill-wright and Miller's Guide, *a standard reference work for more than a century, shows his automated mill. Receiving power from a waterwheel, the machinery took grain from ships or wagons, weighed and cleaned it, then moved it upstairs for temporary storage. Later, conveyors carried the grain to the millstones for grinding, then transported the flour to waiting barrels.*

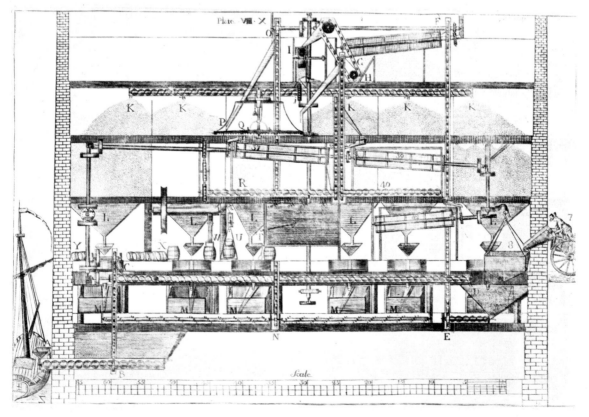

Engines large and small recall the heyday of steam power at a country fair in Virginia. The 4-foot-long, one-fifteenth horsepower "popcorn" engine (top) once turned a popper in a movie theater. Here it powers a machine that sews children's names as souvenirs. The 40-horsepower engine below, built in 1895, drives the saw blade in a shingle cutter.

WILLIAM L. ALLEN, N.G.S. STAFF

pressure of the atmosphere" instead of "the elastic power of steam."

He satisfied himself by experiment that he could make "steam waggons," but he could not make anyone else believe this. And in the 30-plus years that spanned that enlightening Christmas experience and the maiden land voyage of the Orukter, no one was to take him quite seriously enough on the subject.

In his early 20's Evans developed the first of his many inventions, a machine to do the tedious work of cutting, bending, and mounting wire teeth in leather cards used to comb and cleanse wool. This device was so freely pirated that he never made much money from it, though he was always a careful businessman.

Always, too, he was an American patriot. During the War of Independence he joined a militia company—but no fighting came his way.

At the age of 27—rather late for the period—he married, in the spring of 1783; and about this time he devised a number of highly ingenious machines and methods to improve the manufacture of flour in particular and milling in general. His endless belt to lift grain replaced two sturdy men hoisting a tub from the ground floor to the upper-floor granary. His mechanical rake, called a hopper-boy, did the work of one hired youth, spreading the meal evenly to feed it into the machinery at a constant rate.

In later years, although Evans would complain bitterly that an inventor could never get sufficient financial backing, these inventions would support his experiments, himself, his wife Sarah, and, eventually, seven children.

Evans made such a good name for himself as a miller, a manufacturer of milling equipment, and also as an inventor, that President George Washington installed his machinery at Mount Vernon and consulted him as an expert. Justifiably proud of such distinctions, Evans never let business stifle his hopes for steam vehicles.

Opening a store in Philadelphia in 1793, he considered a new and potentially profitable item:

plaster of Paris, then becoming popular for stucco and as lime in farming. But that meant grinding gypsum, slow work for a horse-turned grindstone. Why not use the little engine he was developing for his land carriage?

He started constructing the engine, thinking he could "fully try" the experiment for $1,000. Before he had done, he had expended $3,700—"all that I could command!" Now he had "to begin the world anew" and support his family: "at my advanced age"—he was 48—". . . my way . . . appeared dark and gloomy indeed!"

Happily, this was one of his successes. He boasted of grinding 12 tons of gypsum in 24 hours. To demonstrate his engine "more fully to the public," he set it up in Market Street to drive 12 saws in heavy frames, cutting marble at the rate of 100 feet in 12 hours.

As Evans said, it made "a great show" for thousands of spectators, and he noted with deadpan amusement that they often inquired "if the power could be applied to saw timber, as well as stone. . . ."

Evans's fame as a builder of steam engines came to exceed his reputation as a millwright. When he opened his Mars Works in 1806, orders arrived with welcome frequency. From then until his death in 1819, the works produced almost a hundred engines and boilers.

Well aware of his skill, the Board of Health accepted when he offered to construct his steam dredge for the harbor. He created the Orukter, commonly called "the mud machine." But the scale of this undertaking was so great that he became discouraged once again. Such projects were simply too big for someone who had neither great wealth nor assured backing.

He was by no means free of adversity. In 1805 Congress, which everyone had expected to extend the patent rights to his milling inventions, suddenly and for reasons unknown denied the renewal. This meant, of course, that anyone could use his innovations without paying him any license fee.

Finally, disheartened, Evans wrote to President Jefferson, explaining his troubles: "My crime or rather error is . . . in discovering and making useful inventions by which my country is or will be benefiting to the amount of at least a million of dollars per annum. . . ." Bitterly, Evans declared that "it is really my interest for the good of my family, to withhold from the public, other discoveries. . . ."

Jefferson sent a courteous if not a consoling reply, and promptly signed a bill of relief when Congress passed it in 1808.

Yet only the next year Evans grew so discouraged and overwrought that he took all of his drawings, plans for inventions yet unknown, and after calling his family together he told them that he was going to destroy the drawings "for their good," so that none of them would be tempted to follow in his footsteps, "into the same road to ruin."

"They all approved and I laid them on the fire."

It is rather ironic that in the following years Evans made a good deal of money.

That he was able to take his rebuffs and still work productively is, clearly, in part the result of his harmonious family life. It is also the happy result of his fine sense of humor.

Evans's writings abound with excellent ironic touches. In fact, one might label him the Mark Twain of engineering literature. For such a busy man he did an amazing amount of writing.

In financial straits in 1805, he had to bring out, prematurely, a book on steam engineering that he had been working on for many years. With characteristic wit, but a shade of bitterness, he entitled it *The Abortion of the Young Steam Engineer's Guide.*

Of much greater success was *The Young Millwright and Miller's Guide* that Evans published in 1795. This ran to 15 editions. His biographers, Greville and Dorothy Bathe, knew a mill owner in the 1930's who still consulted it.

The Evans wit and temper appear vividly in a satire of 189 pages, *Patent Right Oppression Exposed, or Knavery Detected,* published under the pseudonym "Patrick N. I. Elisha, Esq. Poet Laureate." This includes an often-quoted prophecy: "The time will come when people will travel in stages moved by steam engines, from one city to another, almost as fast as birds fly, fifteen or twenty miles an hour. . . .

"A carriage will set out from Washington in the morning, the passengers will breakfast at Baltimore, dine at Philadelphia, and sup at New York the same day. . . .

"And it shall come to pass, that the memory of those sordid and wicked wretches who opposed such improvements, will be execrated, by every good man, as they ought to be now."

In Evans's time, the journey by coach from Philadelphia to Boston would mean, say the Bathes, "at least six days of the most excruciating discomfort." Winter, with its snowdrifts and washouts in the roads, could double the time and the difficulties.

It was a different country then. And fortunately for the young American Nation, Oliver Evans was a stubborn man. He fought tenaciously against odds that defeated other talented men.

He had grand dreams. He wanted to see the country full of railroads and turnpikes. Over them would steam the great land carriages that as yet lived only in his mind. And, on his own, he invented the only steam engine light enough and powerful enough for such work: the high-pressure steam engine.

Oliver Evans was a great man and a great American inventor, but in more favorable times he could have been greater and, sadly, he knew it. To one of his sons he left a document called his "Philosophy of Life." It concludes:

"Therefore he that studies and writes on the improvements of the arts and sciences labors to benefit generations yet unborn, for it is not probable that his contemporaries will pay any attention to him, especially those of his relations, friends, and intimates; therefore improvements progress so slowly."

Built to scale, the wooden model below shows the operation of a mill designed by Oliver Evans to grind gypsum for plaster of Paris. Forerunner of automation, the mill took its power from a high-pressure steam engine. Steel hoops reinforced the long wood-sheathed boiler. Gypsum from the wheelbarrow went into the hopper next to it. Broken there by a simple grinding machine, the crushed rock traveled up the enclosed conveyor and dropped into a second hopper for a final grinding. One man operated the mill.

VICTOR R. BOSWELL, JR., N.G.S. STAFF, ATWATER KENT MUSEUM, PHILADELPHIA

ROBERT FULTON and JOHN STEVENS
By Paul Douglas

IN 1792, while the sorry figure of John Fitch wandered around the Philadelphia waterfront, a handsome, dark-eyed, ambitious young man was sipping sherry in the drawing room of an English lord. Although he was an American, the fact that he was a painter with a charming personality redeemed him and he conversed casually with the English aristocracy. Two decades later Americans would honor him as the inventor of the steamboat.

With more precision, a modern writer has said: "Until Robert Fulton, the steamboat would not *stay* invented."

Born the son of a farmer in 1765, Robert Fulton grew up near Lancaster, Pennsylvania, and by the age of 20 had established himself as a painter of miniature portraits in Philadelphia. He was so skillful that Benjamin Franklin asked him to paint his portrait, and no doubt the two men politely discussed politics, the arts, and Franklin's inventions.

An illness in 1786 brought him to the warm springs of Bath, Virginia, where James Rumsey had worked at his mechanical boat. At the resort Fulton may have heard of Rumsey's experiments. Back in Philadelphia the next year he must have heard of, and may even have seen, Fitch's absurd boats on the Delaware.

In 1787 Fulton journeyed to England with a few dollars in his pocket and a letter of introduction from Franklin to another man from Lancaster, Benjamin West, who had become a world-famous painter. Studying under West, Fulton showed some promise; he exhibited two portraits in the Royal Academy in 1791.

In a letter to his mother he told of his past and his prospects: "... I went on for near four years —happily beloved by all who knew me or I had long ear now been Crushed by Poverties Cold wind—and Freezing Rain—till last Summer I was Invited by Lord Courtney [Courtenay] down to his Country seat to paint a picture of him which gave his lordship so much pleasure that he has introduced me to all his Friends. ..."

But the artist had another side. Fulton suddenly gave up painting, and turned to engineering: an art he had to teach himself. His steamboat experiments began in June 1793 with an unusual concept. He wrote: "my first design was to imitate the spring in the tail of a Salmon: for this purpose I supposed a large bow to be wound up by the steam engine and the collected force attached to the end of the paddle . . . to be let off which would urge the vessel forward." He was working on the problem which defeated other inventors—the most efficient means of propulsion. Further work convinced him that steamboats "should be long, narrow and flat at bottom, with a board keel" to reduce resistance.

Fulton's next attempt to achieve wealth and fame came in the form of his submarine, the *Nautilus.* He saw navies in general, and the British navy in particular, as the cause of strife in the world. Thus it seemed logical to seek French support for his project, which he thought would make navies impossible.

He first tested the strange vessel successfully in July 1800, in the river Seine. A few months later he felt ready to blow up British ships in the English Channel. Fulton and his two assistants "remained . . . 6 hours absolutely under water, having for the purpose of taking air only a little tube," waiting for the tide to carry them close enough to the brigs to place explosive charges. One can imagine Fulton and his companions in the dark, dank craft—lighting a candle occasionally to check their watches, their nerves strained and their oxygen scarce. Keeping on the alert, the British spotted the submarine and moved their ships. Thereafter Fulton channeled his energy to the construction of a steamboat.

Analyzing the difference between Fulton and John Fitch, a historian of steamboats, James Flexner, has said: "Fulton's vitality spilled over into other projects. Unlike Fitch, to whom the idea of the steamboat had become the self and the world and God—the whole universe—Fulton was a professional engineer seeking an idea, an

THE NEW-YORK HISTORICAL SOCIETY

Robert Fulton's North River Steamboat of Clermont, *first commercially successful steamboat in the United States, glides at five miles an hour down the Hudson River on a run from Albany to New York City. The inventor enlarged and virtually rebuilt the boat after its initial public trial on August 17, 1807. A skilled artist, Fulton painted his portrait (left) on ivory.*

inventor in search of an invention." Whereas Fitch was an 18th-century artisan with a vision of the future, Fulton was a 19th-century engineer who used precise theoretical calculations before attempting practical experiments.

But in addition to engineering skill, a hopeful steamboat pioneer needed financial and political influence. Fulton found a man with both in Robert R. Livingston, Minister Plenipotentiary of the United States to France.

Not only had Livingston been a member of the committee that drafted the Declaration of Independence, a framer of the constitution of New York, first chancellor of that state, a diplomat and statesman, but also he was a man who saw the value of the steamboat. No longer would steam transportation be advocated by visionaries from the lower class like Fitch and Rumsey; the aristocracy had entered the contest.

Livingston's first serious venture into the field of steamboating had begun in 1797, when he formed a partnership with his brother-in-law,

CARNEGIE LIBRARY OF PITTSBURGH, R. W. JOHNSTON, PHOTOGRAPHER (TOP); LIBRARY OF CONGRESS

Numbers swelling as swiftly as the Mississippi in
flood, stern-wheelers paddled along inland rivers for
more than 150 years. This cluster at the Pittsburgh
quay in 1911 celebrates a century of steam navigation on western rivers. In September 1811 the New
Orleans began the first journey downriver from
Pittsburgh, although no passengers would risk the
trip. (A reproduction of the boat lies just left of
center in the picture above, dwarfed by later models.)
Surviving flood, earthquake, and an attack by Chick-
asaw Indians, the boat, built by Nicholas J. Roosevelt
and Robert Fulton, reached New Orleans in 14 days.
In 1884 one of the Mississippi's last big cotton
carriers, the Pargoud (left), hauls 5,000 bales down
"Old Man River" to the port of New Orleans.

STEVENS INSTITUTE OF TECHNOLOGY
Col. John Stevens (1749-1838)
Pioneer in steamboat and railroad travel

Col. John Stevens, and engine-builder Nicholas Roosevelt, to construct a steamboat. Stevens was a man of some wealth and as influential in New Jersey as Livingston was across the Hudson.

When Fitch was wearing worn and tattered clothing in Philadelphia, Stevens advertised in a local paper that he had lost a trunk containing "Six ruffled shirts, marked J. S.; a pair of black silk smalls, a pair of Buff casimir ditto, a white coat with silver buttons...." The wealth and prestige that these articles represented were indeed, as Rumsey had put it, "a necessary Conection of a Steam boat."

Stevens became interested in steamboats after reading the pamphlets of Fitch and Rumsey. His first plans for a steam engine were similar to those of Rumsey, but his major improvement was in his steam boiler, which allowed better control of the steam, at pressure as high as four atmospheres, and his method of applying the steam to the piston in both ascent and descent. Stevens was proposing a high-pressure engine.

Construction of this engine began in 1791, and Stevens's instructions to his mechanic, John Hall (a "confirmed sot" according to the colonel), indicate that Stevens was at this time a man of ideas rather than an experienced engineer. He told Hall to make a piston of cork, and wheels and cogs of wood. A steam engine of wood and cork! No wonder it failed. Stevens had a lot to learn about these things.

However, he had other interests—the "Villa Stevens," his home at Hoboken, his extensive lands, and his growing family, his plans for a water system in New York, a floating bridge across the Hudson, steam transportation by rail.

But a letter Stevens received in 1795 rekindled his interest in steamboats. John Fitch, his hopes for a steamboat shattered, asked for a loan of 20 dollars and offered four-tenths of the patent rights on a horse-driven ferry. Predictably, things went awry for Fitch. Not only was his request refused, but it also aroused a man who would succeed where he had failed.

Fitch still held the rights to steam navigation on the Hudson, but by 1798 he was virtually forgotten and presumed dead. Livingston managed to get that monopoly repealed and to secure one for himself. The gentleman from New York was losing no time.

The partnership of Livingston, Stevens, and Roosevelt was a strange one. Roosevelt (distant relative of the Presidents) was a dedicated worker and a skilled mechanic; Stevens a theoretician with a strong touch of practicality; and Livingston a headstrong and opinionated man who shot off scores of letters to the patient Stevens and the exasperated Roosevelt, demanding that some new device be tried out.

Livingston's suggestion for a horizontal paddle wheel was tried; it failed. When he demanded that the boat be made as light as possible, it vibrated so much it almost sank. When he departed to assume his duties in France, Roosevelt must have muttered, "Thank God."

Stevens continued to work on the problem and began to build a simplified high-pressure engine. One morning he awoke in bed, his mind excited with a new idea for propelling a boat. Having no paper, he sketched his plan on his sleeping wife's back. When she felt his hand and woke up he asked her, "Do you know what figure I am making?" "Yes, Mr. Stevens," she replied, "the figure of a fool."

On a May day in 1804 a young college student reported that he saw "...a crowd, running toward the river. On inquiring we were informed that Jack Stevens [the inventor's son] was going over to Hoboken in a queer sort of boat. . . . in which there was a small engine *but no visible means of propulsion."*

This boat, the *Little Juliana,* though neither fast nor efficient, was well ahead of her time in having twin-screw propellers similar to those that drive modern ocean liners. The engine was a high-pressure type, but it leaked so much steam that Stevens reverted to one similar to Watt's. Despite American attempts, the British engine was still superior. The problem of combining a reliable and powerful engine with an efficient system of propulsion was yet to be solved.

But a personal matter also bothered Stevens momentarily. He received a letter from Paris in which Livingston told of meeting a promising young inventor named Robert Fulton. Fulton, wrote the former chancellor, ". . . was my partner in an experiment made here on the Steam Boat, and his object is to build one in the United States by way of experiment. . . ." Stevens was no fool. He could read between the lines and see that Livingston had new plans.

Fulton and Livingston each saw in the other a perfect partner. The chancellor had money and the rights to the Hudson; Fulton had the theoretical and practical knowledge of steamboats and the ability to avoid the mistakes of others. By 1802 he had set up a 66-foot-long model tank in France to determine whether "paddles, skulls [sic], endless chains, or water wheels" were most efficient. The problem that plagued Fitch, Rumsey, and others was being resolved in a tank of water! With Fulton's genius and Livingston's monopoly, success seemed inevitable.

Finally, after almost 20 years, Fulton returned to America to see his ideas incorporated in a full-size boat at last. By August of 1807 a new Boulton and Watt engine — its export now legal — was installed and preliminary trials took place. One observer called the boat "an ungainly craft looking precisely like a backwoods sawmill mounted on a scow and set on fire." But Fulton had solved the problem of propulsion by combining an English engine with side paddle wheels on his efficiently designed hull.

The public trial of the boat — first called the "North River Steamboat" and finally named the *Clermont* after Livingston's estate on the Hudson — took place on August 17, 1807. The event rated mention in only one newspaper, the *American Citizen,* which said the boat was "invented with a view to the navigation of the Mississippi upward." On that August day the only hope was that she would reach Albany.

The apprehension of Fulton and his forty well-dressed passengers was thicker than the smoke billowing out of the stack. When the boat finally moved and the creaking and splashing of the side wheels became a reassuring sound, the passengers relaxed and returned the salutes waved by small groups of amazed people who gathered along the banks. On the second day, when the boat reached Clermont, Livingston announced the engagement of his cousin Harriet to the promising inventor. The obscure painter was raised to the social level of his partner. Robert Fulton was a success.

One of the passengers on the first commercial trip in September was warned by a friend not to take such a risk. "John," he was told, "will thee risk thy life in such a concern? I tell thee she is the most fearful wild fowl living, and thy father ought to restrain thee." But others took a chance, and in only two months the boat had cleared 5 percent on the capital expended. Construction of another boat, the first of many, was begun. Although later plagued by lawsuits attacking their monopoly on the Hudson, Livingston and Fulton continued to succeed.

Fitch's boat, although a little faster, was never popular; but those of Fulton, endorsed by the gentry, were filled with passengers who desired comfort and speed on the Albany run.

Now Fulton thoroughly enjoyed the life he had longed for. With an aristocratic wife, a son and three daughters, and an elegant home, he could return to painting not as a struggling young artist, but as a man of leisure enjoying a hobby.

On the other side of the river Colonel Stevens built more boats and argued, with well-bred politeness, with Livingston over the chancellor's monopoly. Livingston must have felt some guilt, for he offered his brother-in-law a share in his partnership with Fulton.

Of course, Stevens refused. By the summer of 1808 he had launched his side-wheeler *Phoenix* and planned to run her on the Delaware, where Fitch had met with indifference. When she traveled from New York to Philadelphia, she made the first sea voyage by a steamboat. The elegant *Phoenix* was a success, and by the time Stevens died in 1838 he saw his plans for steam transportation realized.

With Stevens plying the Delaware and Fulton the Hudson, the age of the steamboat had arrived, and the continued financial success of the Fulton boats tempted other steamboat operators to challenge the monopoly. In *Gibbons* v. *Ogden* the Supreme Court ruled the Hudson River monopoly unconstitutional, opening the rivers of the Nation to steamboats and free enterprise.

In February 1815, while Robert Fulton was inspecting his boats during the winter season, he caught a fatal chill. His funeral was attended by persons of national fame and the shops in New York closed for an hour in respect for the inventor. Even his death was a greater public success than Fitch's.

The national importance of the 1824 decision in *Gibbons* v. *Ogden* is in stark contrast to the obscurity that Fitch, Rumsey, and others had labored in but a few years earlier. The age of the skilled artisan whose invention affected the whole world was waning, and the age of the professional engineer and entrepreneur with impressive financial backing was emerging.

The world would see a few more John Fitches, chugging along with their eyes focused on a vision and their pockets empty, but the new age was one that demanded and rewarded the Stevenses, Livingstons, and Fultons.

SOUTHERN RAILWAY COMPANY (TOP); SOUTHERN PACIFIC COMPANY

Locomotive "West Point" rolls slowly through South Carolina in 1831, its passengers entertained by musicians and protected from possible boiler explosions by bales of cotton strapped onto the "barrier car." Below, two eras meet as pioneer wagon train greets continent-spanning iron horse at Monument Point, near Promontory, Utah, in 1869. Workmen in the United States could lay track rapidly because of several contributions of Robert Stevens, son of steam pioneer Col. John Stevens. Robert invented today's T-rail, wooden ties, and the hook-headed railroad spike, and first used crushed rock for roadbeds.

Agriculture: New Machines for New Lands

3

"CROWN of all other sciences," Thomas Jefferson called agriculture. As a man whose pen could jostle thrones and establish seats of learning, he could speak with authority. As a genius with a flair for practical gadgets, he set out to design a more efficient plow, one that any good woodworker could copy perfectly. In his time, Americans still used wood for many items that Europeans made from metal.

Jefferson was not the only planter interested in better implements. George Washington devised a "barrel plow"—a harrow carrying a barrel with holes burned into it, to drop seed grain at a steady rate—and found it useful on light soils. But he saw no profit in intricate machines for southern plantations. Careless slaves and ignorant overseers, he said, would break them down so quickly that they would "be no longer in use than a mushroom is in existence."

Giant combines reap a golden wheat harvest in Nebraska. Such mechanization has put farmers among the Nation's most efficient workers. In 1800 a farmer worked 344 man-hours to produce 100 bushels of corn; today he need work only four. An agricultural revolution in the 19th century introduced machines powered by animals; another, toward the turn of the century, shifted the work to steam engines.

NATIONAL GEOGRAPHIC PHOTOGRAPHER JAMES P. BLAIR

Gentleman farmer Thomas Jefferson, whose interests knew no horizons, visits a field where slaves test a plow he designed in 1788. It had a "mold board of least resistance"—much as on today's plows—with a "fore-end . . . horizontal to enter under the sod, and the hind end perpendicular to throw it over."

PAINTING BY CHARLES PEALE POLK,
UNIVERSITY OF VIRGINIA, CHARLOTTESVILLE

Thomas Jefferson (1743-1826)
Statesman, architect, and inventor

Thus, from the earliest years of the American Republic, it appeared that the spirit of liberty and the invention of machinery would make a complicated story, not a simple one.

Jefferson's plow never fulfilled his hopes for it, and his country eventually belied his dream of an agricultural economy. Ironically enough, the improvements in farming that he himself did his best to foster helped hasten the coming of industry. Mistrustful of cities, he considered the independent farmer the ideal citizen of a democracy: secure and, in many ways, self-sufficient. Jeffersonian America was an agricultural society, and a prosperous one.

Already, however, the basis and measure of a nation's wealth had begun to change. Not yet conspicuous enough to demand a name, the Industrial Revolution was bringing new sources of strength to Great Britain even as war in the American colonies sapped the older ones. The elaborate and increasingly efficient spinning machines invented by James Hargreaves, Richard Arkwright, Samuel Crompton, and others, would twist national destinies as well as cotton yarns, foreshadowing the age when a country without industry is poor almost by definition.

So, after Eli Whitney invented his cotton gin, southern planters would grow rich, virtually doubling their cotton crop every decade for 40 years, to supply the mills of England—and of New England. British law made it a crime to export machinery or plans, but in 1789 an immigrant named Samuel Slater reached Philadelphia with the complexities of textile machines in his head. From one small "Manufactory" that he managed in Pawtucket, Rhode Island, grew a whole Yankee system.

To feed an industrial society, with its growing towns and cities, might seem impossible for farmers plagued with much work and few hands to do it. Inventions met the need.

In 1837 John Deere introduced his famous plow, "the plow that broke the plains," turning the tough prairie sod and polishing its own

"All my wishes end where I hope my days will end," wrote Jefferson of his Virginia estate, Monticello. He designed the house and filled it with gadgets like the candle-lighted swivel chair in motion below.

NATIONAL GEOGRAPHIC PHOTOGRAPHER DEAN CONGER

NATIONAL GEOGRAPHIC PHOTOGRAPHER DICK DURRANCE II

ARTHUR LIDOV

tempered steel in the process. Reapers—and later combine harvesters—clanked into the fields, and in the 1880's steam traction engines began to replace straining teams of horses. Improved hybrid plants and commercial fertilizers increased production still more.

In 1800 a farmer expended 344 man-hours to raise 100 bushels of corn; in 1910 he worked 147 for the same amount; by 1960, only 4. Before 1870 one farm worker produced enough food and fiber for five townspeople; by 1900, for seven; by today, for more than 30. In Jefferson's time 70 percent of the labor force of the United States was employed on farms; today that figure has shrunk to less than 7 percent.

In this revolution of abundance, old distinctions between agriculture and industry blur and fade. The invention designed for the one may serve the other as well.

So it turned out with the sugar-refining equipment of one of the most neglected of major American inventors—Norbert Rillieux, born in New Orleans in 1806, legally described in Louisiana records as a "quadroon libre" or "free man of color."

His father (who was white, an engineer and inventor himself) gave him his own surname and an excellent education in Paris. There young M. Rillieux learned steam engineering, which he applied brilliantly in designing his multiple-effect evaporator.

This system, first patented in 1843, transformed cane-juice processing, sharply reducing costs, producing purer sugar with greatly increased efficiency. It is used today in any industry (such as soapmaking) where thin solutions in quantity must be processed cheaply.

Rillieux won respect and a comfortable fortune in Louisiana, but these could not spare him social and legal indignities. As the Civil War swept his country, he returned to France. Jefferson's most enduring dream, of liberty for all, had not fully come true by 1894, when Norbert Rillieux died in Paris.

Harvest fire strips foliage from sugarcane in Hawaii. Later, the cane will yield its juice in a multiple-effect evaporator, an apparatus invented in 1843 by Norbert Rillieux (lower). Described in a patent application, his evaporator consisted of a series of vacuum chambers, connected to "make use of the vapor of the evaporation of the juice in the first, to heat the juice in the second," and so on. An evaporator at the Sugar Cane Growers Cooperative of Florida at Belle Glade (lower left) employs the same principle today.

CATHEDRAL HOME
SCHOOL

NATIONAL GEOGRAPHIC PHOTOGRAPHER JAMES L. AMOS (ABOVE); GRANT AVERILL (LOWER LEFT); ARTHUR LIDOV

ELI WHITNEY
By Edwin A. Battison

"THE ARTIST of his Country," a contemporary called Eli Whitney in 1801, adding, "all Men of all parties agree that his talents are of immense importance. . . ." But although renown had begun to temper the disappointments of his early career, Whitney and his backers still faced years of struggle and frustration.

First child of Eli Whitney, Sr., and Elizabeth Fay, Eli was born on December 8, 1765, and spent his youth on the family farm near Westborough, Massachusetts. His mother, chronically ill for years, died when he was 12. A stepmother and two stepsisters, acquired two years later, earned his dutiful respect, but no more.

Eli thus grew toward manhood in a home with family tensions, in a country embroiled in revolution. The shot heard round the world missed the Whitney farm by a scant 12 miles. War brought hard times and, incidentally, began Eli Whitney's manufacturing career.

Shortages drove up the price of everything, including nails, which had usually been imported. Eli, now 14 and mechanically inclined—he had made a creditable violin at 12—persuaded his father to let him install a forge in the farm workshop. He began turning out and selling nails. When demand exceeded supply he decided to expand. Unbeknown to his father, he saddled up and rode off, looking for a workman to help him. He found his man and visited many shops, probably picking up pointers on the mechanic arts. His first view of the world seems to have been from this trip.

Peace in 1783 brought an influx of foreign goods—including cheap nails from England—and Eli's business dropped. He switched to hatpins for ladies and to walking canes.

When nearly 18, Eli professed a yearning to attend college. As a first step, he became a schoolteacher. In the days of ungraded classes anyone who satisfied the town authorities could have the job, and it gave Eli good training. In order to teach, he had first to learn, studying diligently to stay ahead of his pupils. For five winters he taught; during the summers he attended an academy in Leicester.

At 23—a very mature age for that period—over his stepmother's objections but with the promise of money from his father, Eli departed for New Haven, Connecticut, and Yale College.

His years at Yale proved invaluable to him. Intellectual ferment stimulated his already active mind, companions from other parts of the country broadened his horizons, and social intercourse smoothed his country-boy manners.

More important, however, were the contacts he made. Throughout his life—especially at moments of crisis—Yale graduates appeared with a helping hand, often in positions of importance; he became adept at finding them.

Although he worked to help support himself, he had to write to his father: "I have succeeded very well in my studies, & meet with no other difficulties but the want of money, which indeed is very great."

Commencement exercises in September 1792 turned Eli out into the world, penniless and nearly 27 years old. Intending to study law, he engaged himself to tutor the children of a Major Dupont, a South Carolina planter. He embarked for the South in the company of some new friends. Catherine Greene, widow of the late Gen. Nathaniel Greene, her children, and her estate manager, Phineas Miller, were returning to Georgia after a northern visit, and Eli, recovering from inoculation for smallpox, sailed with them.

Approaching her forties, Mrs. Greene was a vivacious person, generous, famed for her devotion to her husband and the patriot cause. At Mulberry Grove, the plantation given to the general by the State of Georgia, Eli heard that he might get only half the promised salary for his tutoring. He accepted Mrs. Greene's invitation to stay as her guest.

Within two weeks of his arrival in 1793 he had invented the machine that made his fame.

Cotton growing in the South was in its infancy. Two varieties were cultivated: a fine,

YALE UNIVERSITY ART GALLERY, GIFT OF GEORGE HOADLY, YALE 1801

Yankee tinkerer Eli Whitney gazes from a portrait painted by his friend Samuel F. B. Morse in 1822. Whitney transformed the South with his cotton gin, invented in Georgia in 1792. The patent drawing in cross section below illustrates its simplicity: Slender wire fingers, picking up cotton fed into the curved hopper, carry the fibers through the grid at top. Too fine for seeds, the grid lets only cotton pass through; wire brushes clean the spikes. Southerners found the gin easy to pirate, and Whitney spent years fighting for a share of the wealth from cotton.

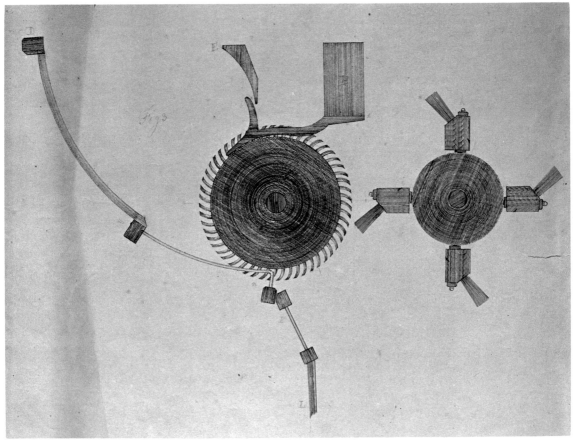

ELI WHITNEY PAPERS, YALE UNIVERSITY LIBRARY

Early cotton gin (below), invented before Whitney's, removes seeds only from a strain of cotton grown near the coast. Whitney's machine (opposite, lower) could gin any variety, including one that thrived inland. Within a few years after his invention of the new gin in 1792, the cotton crop increased phenomenally: In 1795 planters grew eight million pounds; 12 years later the crop had jumped to 80 million pounds. At right stands a market-bound wagon.

COURTESY JAY P. ALTMAYER (UPPER); HARPER'S WEEKLY, 1869, LIBRARY OF CONGRESS

Patent model of Whitney's invention survives at the Smithsonian Institution. Whitney planned to set up gins throughout the South where planters would bring crops for processing, but high fees, the ease of copying the Whitney machine, and a fire in his Connecticut factory prompted planters to build their own gins. The resourceful Whitney turned to manufacturing muskets for the Government.

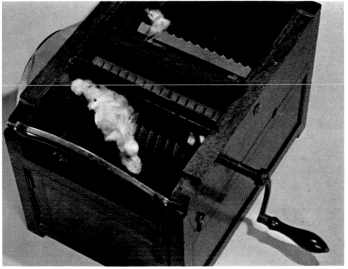

VICTOR R. BOSWELL, JR., N.G.S. STAFF, SMITHSONIAN INSTITUTION

Mechanized cotton picker, lumbering symbol of an ever-changing agricultural technology, works a field near Scott, Mississippi. Such machines relieve pickers like the one below of long drudgery in the southern sun—but increasingly mechanization relieves them of their jobs as well. A two-row picker costs about $22,000 and can do the work of 120 field hands, gathering 4,500 pounds of seed cotton an hour.

delicate strain, black-seed or sea-island cotton, would grow only along the coast; green-seed cotton grew well in the interior, but its fiber clung tenaciously to the seeds. Picking seed from a pound of cotton by hand was a day's work for a nimble-fingered woman or youth.

A ready market existed in England for any cotton grown. In addition, a smaller market was developing in New England. Vast areas of the South were still wilderness, awaiting cultivation. A golden future awaited; all that was wanting was a key to unlock it. Eli Whitney wandered in and, almost offhandedly, provided it.

Planters visiting Mrs. Greene had talked of the problem; she urged him to provide a solution. His mind "involuntarily" running on the subject, he wrote, he "struck out a plan. . . . In about ten Days I made a little model. . . ."

A gin for black-seed cotton already existed: The fibers passed between two rollers and the smooth seeds dropped out. But the fuzzier green seeds would pass right on through, mashed into the fibers.

In Eli's machine, wire teeth on a roller tugged the fibers through slits in an iron guard—slits too narrow for the seeds to pass through. "The cotton is put into the Hopper," he wrote, "carried thro' the Breastwork by the teeth, brushed off from the teeth by the Clearer and flies off from the Clearer with the assistance of the air, by its own centrifugal force. The machine is turned by water, horses or in any other way as is most convenient." It was simplicity itself, and its very simplicity would cause its inventor hardship and sorrow for years to come.

Spurred by visions of untold riches, Whitney formed a partnership with Miller, a fellow Yale man. The partners promptly made two mistakes: They failed to guard the model from sight and hearsay, thus inviting piracy; and they let greed clog their judgment. Instead of selling the gins outright or selling the right to manufacture them, they planned to establish gins at strategic locations and charge two-thirds of a pound of cotton

NATIONAL GEOGRAPHIC PHOTOGRAPHER JAMES P. BLAIR

for every pound returned to the planter. This was an exorbitant rate.

Moreover, they announced their readiness for business too soon. While Whitney was in Philadelphia applying for a patent from Secretary of State Thomas Jefferson—who wondered if he might buy a gin—Miller put a notice in the March 6, 1794, *Gazette of the State of Georgia:* "The subscriber will engage to gin . . . any quantity of the green seed cotton. . . . ginning machines . . . will actually be erected . . . before the harvest of the ensuing crop."

Before the harvest of the ensuing crop? With a little luck, they might have made it.

In New Haven, Whitney had a shop that he had used to construct his patent model. There he began building gins, worried that "we shall not be able to get machines made as fast as we shall want them." By the harvest they had a couple functioning at Mulberry Grove and a factory— a "Barn 26 feet by 20"—built in New Haven.

Prospects were bright as 1794 ended. Planters were committed to growing more green-seed cotton; Miller had arranged credit—with Mrs. Greene's help—and was touring the South buying sites; the new factory was functioning.

Then, catastrophe. On March 11 fire left the new building "all in ashes! My shop, all my tools, material, and work equal to twenty finished cotton machines all gone—"

The planters' crops could not wait for Miller & Whitney to recover. From neighbors who had seen the first model, descriptions spread. Any mechanic, once he grasped the idea, could easily make one. Soon the South was dotted with humming cotton gins, and all the lawsuits in the world couldn't stop them.

For years Miller & Whitney tried. In repeated trips through the South, Whitney pleaded for redress from state legislatures: His invention was making men rich in the South and he was not making a penny. The states should, therefore, pay him a flat fee for the right to use his machine. In December 1801 South Carolina

authorized payment of $50,000. Miller believed Whitney's polite insistence had increased it, "for without his presence I should have feared that five and twenty thousand alone would have been obtainable." Other states eventually took similar action, but Whitney barely covered expenses before his patent expired in 1807.

By 1811 he was embittered. He wrote to Robert Fulton: ". . . this Machine being immensely profitable to almost every individual in the Country all were interested in trespassing, & each justified & kept the other in countenance."

Others paid more dearly. Planters came to believe that slavery was an economic necessity. Its grip would not be loosened for half a century. In the North, grim mill towns would grow up. In English cotton factories of the 1830's children often worked 12 to 14 hours a day; in some mills relentless overseers drove them past the point of exhaustion.

But the era of reform was yet to come, and Whitney had formed a new idea for a factory of his own. By 1798, when prospects for a profit from the gin seemed darkest, he had coolly appraised his situation. He saw that inadequate financing had hampered him, and turned to one backer that could certainly support experimental enterprises: the Federal Government.

"Bankruptcy & ruin were constantly staring me in the face & disappointment trip'd me up every step I attempted to take," he wrote, but on June 14, 1798, he signed a contract to manufacture 10,000 complete stands of arms—muskets with bayonets and ramrods—for $134,000. (If $13.40 seems cheap for a gun, Whitney had on one occasion paid $12.75 for 17 days' board in New York, and that probably included his room.)

Whitney had a little more than two years to fulfill his contract: incredible confidence! But he had ideas on how to accomplish it.

"Manufacture" still meant "to make by hand." In an armory, for instance, a workman hammered, filed, carved out, and fitted a musket. And then he started over and made another.

Industrious to the end, Whitney made his last sketch a few days before his death on January 8, 1825. It suggests a tool for making a part for musket locks. He based the idea on three drawings of similar machines already in use by rival contractors.

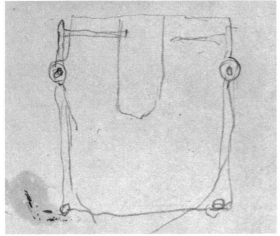

ELI WHITNEY PAPERS, YALE UNIVERSITY LIBRARY

In France, an inventive mechanic named Honoré Blanc had developed a system of interchangeable parts for gun-making. Jefferson had visited his shop in the 1780's, seen parts "gaged and made by Machinery," and tried in vain to persuade him to move to the United States.

Whitney had almost certainly heard of this from Jefferson. But with all the innovations imaginable he couldn't possibly meet his contract. A $5,000 advance solved the problem of his solvency, but he had no factory, no trained workmen, no materials, no new machines.

If Whitney originally doubted his ability to deliver the arms on time, he concealed this. He knew just what he meant to do; but nothing in his experience with the cotton gin, a simple and lucky invention, prepared him for the tasks of accurate production he was attempting.

The Government's reasons for agreeing to such an unrealistic contract included its respect for Whitney as an inventor; its own ignorance of industrial techniques; its determination, as a Congressman said, to save the Nation from "a disgraceful recourse to foreign Markets for this primary means of defence." It would pay $4.40 apiece more for Whitney's guns than for those imported from Europe.

Whitney took more than ten years to fill that first contract. Routine construction met unforeseen delays. New concepts were not easily carried out: "One of my primary objects is to form the tools so the tools themselves shall fashion the work...."

During the years of the cotton-gin crises, when he was fighting for reimbursement from the southern states, he also made trip after trip to Washington, standing off officers who wanted arms at once, doggedly accounting for his delays and pleading his case with desperate and persuasive urgency—if he lost this contract he would be more hopelessly in debt than ever.

Today, tooling up is familiar to everyone: It takes longer to get ready to manufacture something than to manufacture it. In Whitney's day it was all new and strange.

The officials—Yale men among them—granted him one extension after another. Whitney struggled along, alone, with an unsatisfied yearning for kith and kin that he felt more clearly than he could define. He surrounded himself with his sister's sons, and his 10-year-old nephew Philos Blake described the factory in a letter to his sister: "...a drilling machine and a boureing machine to bour berels and a screw machine and too great large buildings...."

Although Philos did not mention interchangeable parts, they fascinated Government men when Whitney spoke of them; and once, in 1801, he staged a convincing demonstration for Federal officials—he showed them a musket on which he could fit any of several locks. But these locks must have been especially prepared for the purpose; Whitney muskets in several collections —including the Smithsonian Institution's—have parts that cannot be interchanged.

WILLIAM G. MUNSON, "ELI WHITNEY GUN FACTORY," YALE UNIVERSITY ART GALLERY, MABEL BRADY GARVAN COLLECTION (BELOW);
VICTOR R. BOSWELL, JR., N.G.S. STAFF, COURTESY EDWIN A. BATTISON, SMITHSONIAN INSTITUTION (ABOVE); ELI WHITNEY PAPERS, YALE UNIVERSITY LIBRARY

Connecticut's Mill River meanders through Whitney-ville, the village that grew up around Eli Whitney's musket factory. In 1798 the inventor signed an impossible contract with the Federal Government to manufacture 10,000 muskets in a little over two years. Only then did he select a factory site; the contract ultimately took more than ten years to fill.

Right up to the last guns Whitney made, the parts of each lock carried identifying marks that truly standardized parts do not need. In the end, he made his thousands of muskets with the same old differences between them that had plagued armies for generations.

It was left for others, notably Simeon North and John H. Hall, to make enduring contributions to the mass production of interchangeable parts. And late in Whitney's career younger competitors were adopting more advanced methods than his. His renown has overshadowed their work and obscured the fact that he never really reached his goal.

Like so many other dedicated inventors, Whitney sacrificed his personal life for his work. He developed a warm friendship with the charming Mrs. Greene, and after his partner Miller married her he began referring to himself in letters as a "solitary *Old Bachelor.*"

After both Catherine and Phineas Miller had died, Whitney finally married. In 1817, at 51, he took to wife 31-year-old, aristocratic Henrietta F. Edwards, granddaughter of the famed preacher and philosopher Jonathan Edwards. Their marriage was apparently a happy one, though within five years his health was "but indifferent."

As his condition deteriorated, he turned more and more of the operation of the armory over to his nephews. An enlarged prostate caused him agonizing pain; he studied anatomy with his doctor and devised instruments that gave him temporary relief; after much suffering he died on January 8, 1825. His nephews modernized his armory and carried on the business until his son, Eli, was old enough to take over.

An early biographer spoke eloquently of Eli Whitney's contribution to America: "Every cotton garment bears the impress of his genius, and the ships that transported it across the waters were the heralds of his fame, and the cities that have risen to opulence by the cotton trade, must attribute no small share of their prosperity to the inventor of the Cotton Gin."

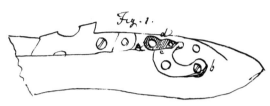

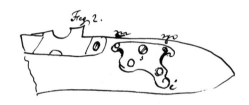

"A good Musket is a complicated engine," Whitney wrote, applying for a second Government contract. "Each musket, with Bayonet, consists of fifty distinct parts." Whitney and others tried for years to manufacture interchangeable musket parts but never quite succeeded; machine tools of the time allowed too many variations. The sketch above shows an idea —never incorporated—for strengthening the lock.

ROBERT W. MADDEN

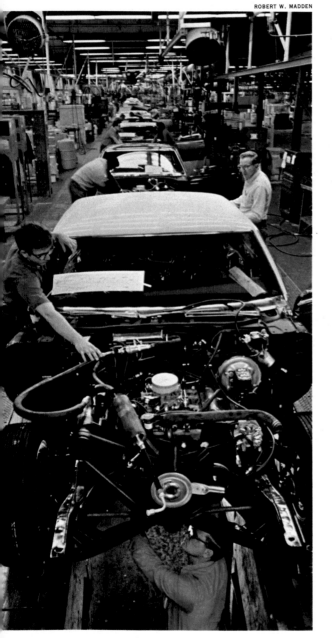

Sparks fly as a molten torrent tumbles into molds at Sparrow's Point, Maryland, in a mill belonging to the Bethlehem Steel Company, a major supplier for the automobile industry. In the first half of the 19th century, the growing network of railroads and new factories created a huge market for steel, but not until the 1850's could American mills meet the demand. After the Civil War, however, America came to rival England as the world's leading steel supplier.

Unfinished Oldsmobiles (left) roll along an assembly line in Lansing, Michigan. An outgrowth of experiments with interchangeable parts and mass-production techniques, this line turns out 95 cars an hour. Ransom Olds first applied the system to building automobiles in 1899, nearly ten years before Henry Ford installed an assembly line. By 1904 Oldsmobile production had reached 5,000 cars a year.

Custom-made car nears the end of an Oldsmobile assembly line. Computers guide and time the components—grilles, instrument panels, engines, and optionals—so they reach the right place at the right time. Fenders for a specific car arrive from sub-assembly lines at just the proper moment. The next automobile in line may look entirely different.

Cyrus Hall McCormick (1809-1884)
Innovative industrialist

By Ronald M. Fisher

HARVESTING WHEAT with scythe in hand is one of those cruel jokes nature reserves for farmers. It is backbreaking, arm-wrenching work. Under a blistering sun, sweat streams into the eyes, biting insects buzz around the head, chaff finds its way down the back of the neck. The wind-blown wheat seems to nod in derision as it marches away toward the horizon.

Roman farmers harvested their grain in much the same manner. Flemish farmers in the 15th century added the cradle, a device for laying the severed stalks evenly, ready to be gathered. This made the scythe even heavier, and generations of farmers have cursed it.

Moreover, ripe wheat demands attention — within ten days, weather permitting. Gathered too late, the grain falls to the ground. Before the invention of the reaper, farmers helped one another with the harvest, but only so much could be done. With a scythe and cradle, a strong and skilled man could harvest two or three acres in a day. It simply was not enough.

In 1838, a notably bad year, America imported grain from Europe. This, despite enormous plains of potential wheatfields. With horse-drawn plows, harrows, and planters, farmers could now sow far more than they could reap.

At Walnut Grove, Virginia, one young farmer cursed the scythe and vowed to do something about it — Cyrus Hall McCormick, born in 1809, of Scotch-Irish descent. His grandfather had moved to Virginia from Pennsylvania and settled there; his father, Robert, had prospered. By 1830 Robert owned some 1,200 acres, two grist mills, two sawmills, a lime kiln, a distillery, and a blacksmith shop. Here he had tinkered for years with ideas for a mechanical harvester. He had many problems to solve.

No reaper could duplicate the swinging motion of a man with a scythe. Instead, the wheat must somehow be cut with a knife. It would have to be cut gently or the grain would fall to the ground. Wet stalks would cling together, snarled and tangled. Rocky, hilly fields sabotaged every effort at mechanical harvesting.

As young Cyrus grew older, he became less interested in being with his five brothers and sisters and more interested in his father's blacksmith shop. His fastidious attention to his clothes, his reserve and earnestness were already occasioning some criticism from his neighbors and alienating him slightly from his siblings.

Like most Virginia farmboys of the time, he had informal and erratic schooling. He attended classes for an unknown number of years in an old log schoolhouse, and was taught surveying and mathematics for at least one year by a tutor. But working with the forge and tools fascinated him, and alongside his father he grappled with the problems of the reaper.

England had granted her first reaper patent in 1799. British inventors continued to try to improve the machines, but farm laborers, desperately afraid of losing work, opposed them stubbornly and sometimes violently.

Green hills of Virginia shelter cloud-shadowed Walnut Grove farm, birthplace of Cyrus McCormick, inventor of the reaper. His father Robert, who tried for years to construct a workable harvester, owned the mill and blacksmith shop below. There, in 1831 at age 22, Cyrus built his first successful reaper.

WILLIAM L. ALLEN, N.G.S. STAFF (ABOVE); CAROL ALLEN (BELOW); NATIONAL CYCLOPEDIA OF AMERICAN BIOGRAPHY AND WASHINGTON AND LEE UNIVERSITY, LEXINGTON, VIRGINIA

In the United States several men patented reapers of one kind or another during the 1830's. Men like Hazard Knowles in the District of Columbia and Samuel Lane of Maine, Hiram Moore and John Hascall of Michigan, Alfred Churchill of Illinois—all built reapers drawn or pushed by horses, but their machines were too few and too faulty.

Eventually, the elder McCormick gave up on reapers, but Cyrus was determined to succeed. By the harvest of 1831, when he was 22, he had a machine that seemed to work. At a demonstration in a field of ripe oats on a neighboring farm, the machine performed the way it was supposed to. Noisy, awkward, and ugly, it flapped and rattled across the field, frightening the horse but cutting grain. Jo Anderson and Anthony, two of the McCormicks' Negro slaves, were there to hold the horse and rake the oats. Within hours the reaper harvested as much grain as two or three men could cut by hand in a day.

The machine combined all of the necessary elements, none of them new, but never before assembled so practically. These included the knife, iron fingers to hold the grain against the knife, a large reel that pushed the stalks back against it, a platform for catching the severed stalks. Also, a heavy wheel directly behind the horse to carry most of the machinery and provide power for the reel and cutter, the cutter mounted to one side so the horse walked in stubble, and a divider at the end of the cutter bar to separate grain to be cut from that left standing.

Other men's reapers had one or several of these elements, but apparently none had them all. In 1834 Cyrus applied for and received a patent for his machine.

But now, with success so near, he seemed to lose enthusiasm for his invention. With his father, he became involved in an iron-smelting venture with ore from a local deposit. The family sank most of its capital into the project. From somewhere Cyrus had scraped together $6,000, but just as they were getting into production the Panic of 1837 hit. The price of iron fell from $50 to $40 to $25 a ton and the McCormicks were plunged into debt. Fortunately, a good friend with good credit was able to provide a loan.

These years of trouble—the darkest of Cyrus's life—forged his character as early success never could have. He discovered the will to fight. He found that repaying the family's debts was more gratifying than being an ironmonger.

It was the reaper that saved them.

Turning to it in desperation, Cyrus built three and sold two in 1840. In 1842 he sold six or seven, and twenty-nine in 1843. The more he sold, the more free advertising he received. Editors, eager for news, publicized them gladly. In 1844 Cyrus built and sold some 50 machines, all made in the blacksmith shop at Walnut Grove and sold for about $100 apiece. By 1847 the McCormicks were out of debt. Cyrus's career was launched.

In 1844 he had taken a trip and discovered something that would change his life. Traveling through the western states—Ohio, Michigan, Illinois, Wisconsin, Indiana, Missouri—he found vast prairies waiting for the plow. Here, he knew, would be the wheatfields of the future.

Here, too, he invaded the territory of the man who would be his first great rival. Obed Hussey was a one-eyed sailor from Maine, a Quaker, and a builder of reapers. He had secured a patent in 1833, six months before McCormick—and had been making his machines in Cincinnati. Though simpler and sturdier than McCormick's, Hussey's version was actually more of a mower.

The two rivals soon tangled. Their combat generally took the form of reaping contests. A field of grain would be selected, the duelists would arrive, each with his partisans, and the race would be on. These contests never proved anything—the conditions varied too much—but the farmers enjoyed them.

Obed Hussey just might have achieved the success of Cyrus had he not made a wrong decision. He turned his back on the frontier, moved to Baltimore, and concentrated on selling to farmers east of the Appalachians. There glory passed him by, although his invention is the prototype of the modern mowing machine.

In 1848 Cyrus settled in Chicago, a town on Lake Michigan just beginning to boom. It had fewer than 17,000 people in 1847, but it would gain 10,000 in three years. Immigrants were pouring in and muddy streets lined with ugly frame houses were sprawling outward.

Obviously Chicago was going to be an important city. Much of the capital of the west was centered here. A trade route to the Atlantic existed via the Illinois & Michigan Canal, the Mississippi River was accessible by railroad, other railroads were pushing into the west.

In the year he arrived Cyrus started a factory, one of the largest in the city. From now on he would invent no more. He would concentrate on improving his reaper—mostly, in later years, by adapting the inventions of others.

But he was only beginning as a businessman.

*Retired blacksmith Harry Wilson, who traces his
ancestry to a McCormick slave named Jo Anderson,
displays his skill in the smithy at Walnut Grove. Here
McCormick—advised by his father and aided by Jo—
toiled through long winters developing the reaper.
At public demonstrations, Jo walked alongside the
harvester, raking the platform clear of severed grain.*

VICTOR R. BOSWELL, JR., NATIONAL GEOGRAPHIC STAFF

Quiet revolution sweeps the Midwest with the introduction of the horse-drawn reaper. In the photograph below—taken as the 1931 centennial of the McCormick reaper approached—harvesters demonstrate the earliest model. As one man rakes the stalks off the platform, others gather and stack them. Besides speeding the harvest, the reaper freed many men. After the Civil War, Secretary of War Edwin M. Stanton said: "The reaper is to the North what slavery is to the South. By taking the place of regiments of young men in western fields, it released them to do battle." Before its invention, men cut grain with scythe and cradle (left) that allowed the stalks to fall with the heads all pointed the same way, ready for binding.

INTERNATIONAL HARVESTER COMPANY (ABOVE); LIBRARY OF CONGRESS

Skeptical neighbor stoops to examine the evidence—mowed oat stalks—at the first public demonstration of the reaper at Steeles Tavern, Virginia, in July 1831. Once in business, McCormick pioneered modern marketing techniques, advertising heavily, sending out agents, and selling on the installment plan.

Over the next few years he developed selling techniques that helped revolutionize business. He began by issuing a written guarantee for his reapers, a novel practice. He sold his machines at a fixed, published price. He invested heavily in advertising, buying space in newspapers and magazines and printing leaflets with detailed explanations of how the machine worked and testimonials from satisfied customers.

He hired agents to represent different territories. He began selling on the installment plan. He promoted field tests, letting everyone get a good look at his machines.

And he sold only machines built in the Chicago factory. Earlier he had hired subcontractors, but inferior versions reached the public and caused considerable hostility toward the McCormick name until the Chicago factory could handle all of the production.

Business boomed. By 1848 Cyrus had sold 1,300 reapers. Between 1848 and 1850 he sold 4,000 more.

The Crystal Palace Exhibition in London in 1851 provided favorable publicity. At first *The Times* was dubious about McCormick's reaper, describing it as "a cross between a flying machine, a wheelbarrow and an Astley chariot [a type of carriage]." But in a public demonstration the "Virginia Reaper" cut wheat at the rate of 20 acres a day. *The Times* admitted that the reaper alone was worth more to England than the whole cost of the exposition. Cyrus won a medal, and became a world celebrity.

During these years he seemed happiest when in court. They were days of free-wheeling charges and countercharges, when rivals hurled suits, complaints, writs, and injunctions endlessly at one another.

After Hussey, Cyrus's most important rival was John H. Manny, who made reapers in a factory at Rockford, Illinois. Cyrus sued him in 1854, alleging patent infringements. McCormick's other rivals helped finance Manny's case in a final effort to check Cyrus's growing power.

Both sides assembled a battery of high-powered attorneys. As the case was to be tried in Springfield, Manny's counsel shrewdly included a local lawyer—Abraham Lincoln. The case was transferred to Cincinnati but Lincoln came along, bringing a meticulous brief that had cost him considerable time and effort. One of Manny's chief lawyers was Edwin M. Stanton, who maneuvered Lincoln into the role of silent spectator. Manny won the case, McCormick hired Stanton for his next lawsuit, and Lincoln, later, would make Stanton his Secretary of War.

Not until 1858 did Chicago's most eligible bachelor take a wife. Almost 49, portly, bearded, and an earnest Presbyterian, McCormick until now had had little interest in anything but his business. He enjoyed certain luxuries—custom tailoring, lavish displays of flowers, good music, travel—and now he asked young Nancy (Miss Nettie) Fowler to share them. The Chicago *Daily Press* could not resist commenting that "in *reaping* one of the fairest flowers our city can boast, he has but added the orange blossoms to the laurels of his world famous title of nobility."

A humorless man, Cyrus would go to extreme lengths to carry a point if he thought himself in the right. In 1862, returning home by train from Washington with an entourage that included his wife and two of the five children they eventually would have, Cyrus found he had been charged an extra $8.70 for excess luggage. Furious, he refused to pay. He dragged his family off the train just as it pulled out, carrying all the luggage with it. Then he contacted the president of the railroad, demanding fair play.

The president issued instructions for the baggage to be held in Pittsburgh, but they went unheeded. The luggage went on to Chicago, where it was stored in the baggageroom. That night lightning struck the station and it burned to the ground. Cyrus was livid.

He sued the New York Central and won. The railroad appealed. The case was tried four times and appealed five, Cyrus consistently winning.

INTERNATIONAL HARVESTER COMPANY (ABOVE); ARTHUR LIDOV

McCORMICK MUSEUM, STEELES TAVERN, VIRGINIA (BELOW); THE BETTMANN ARCHIVE

*Camels pull an early McCormick reaper in a field
in southern Russia. Displayed at the Crystal Palace
Exhibition in London in 1851, the reaper spread
quickly across the Continent, reaching Austria-
Hungary, the German states, and Russia before 1861.*

Cutting, threshing, winnowing, and bagging grain, a ponderous combine works an Oregon wheatfield in the 1880's. The machine, a descendant of Cyrus McCormick's mechanical reaper, clearly indicated the need for engines as a compact source of power.

Tractors undergo a final check at the International Harvester Company in Chicago. The firm grew from a merger in 1902 between McCormick's company and four other farm-implement manufacturers. Cyrus's boyhood home displays a model of his original reaper (left), a crucial tool in developing the West and feeding the world's hungry.

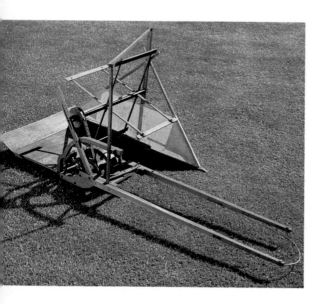

Pennsylvania. Judging from the silence of this correspondence, there was no Emancipation Proclamation, . . . Lee did not surrender, and Lincoln was never assassinated. Business did not go on as usual, but it was all-absorbing.''

During Cyrus's later years, his brother Leander —long a supervisor in the Chicago factory— openly tried to disprove Cyrus's claim to be the inventor of the reaper. The two brothers had squabbled frequently over business matters. Prompted perhaps by jealousy, Leander had been collecting evidence from friends, relatives, and elderly Virginians to prove that their father, Robert, was the true inventor. He failed to convince anyone but himself and his circle, and today historians accept the evidence supporting Cyrus McCormick's fame.

Cyrus died in 1884. His company, 18 years later, merged with four others to become the International Harvester Company, still a giant in the industry.

Cyrus McCormick's great contribution to invention came in a careful, persistent application and refinement of technology. Devising his reaper was an important achievement, but the marketing of it was an even larger factor in his success. He is a prime example of the inventor as businessman, making his product pay.

When he sought the extension of a patent in 1858, Cyrus received testimonials from western farmers. They agreed that the reaper was worth ten times its cost; that it had allowed farmers to double their wheat acreage; that many farmers were turning from corn to small grain because of it. It had helped to make farmers mechanically minded. It encouraged invention of other agricultural machinery. It helped solve the labor shortage. It wasted less grain than the scythe and cradle. And McCormick's easy-credit system permitted farmers to enlarge their farms even as they paid their debts with the return from increased crops.

In short, after Cyrus McCormick, the West began to pay for itself.

Finally, a year after his death, his estate was awarded more than $18,000.

In his later years, Cyrus gave more and more time to his faith. ''Business is not inconsistent with Christianity,'' he wrote. He looked upon his reaper as a blessing for the tillers of the soil.

During the Civil War his sympathies lay generally with the South, but his reaper helped the North to victory. As Eli Whitney—a Yankee— had propped up the South's slave-based economy, so Cyrus—a Virginian—freed thousands of Northern boys from the necessity of returning home at harvest time.

McCormick detested Lincoln's policies and was deeply distressed by the war, but allowed nothing to influence the conduct of business. His biographer William T. Hutchinson summarizes the letters mailed by the McCormick office between 1861 and 1865: ''Fredericksburg signified that the premium on gold might rise, and Lee's march to Gettysburg that fewer reaper sales might be expected in Maryland and

NATIONAL GEOGRAPHIC PHOTOGRAPHER JAMES P. BLAIR (ABOVE); VICTOR R. BOSWELL, JR., N.G.S. STAFF

Legend in the Making: Yankee Inventiveness

4

"THE MOMENT an American hears the word 'invention,'" observed a visitor in the late 1820's, "he pricks up his ears." This alert optimism marked a pervading change from the conservative skepticism of the 18th century, when the word "innovation" had provoked distrust on both sides of the Atlantic.

Now, with pioneers pushing westward, national wealth and power rising visibly, and reformers preaching new ways to perfection, Americans cherished a restless dream of progress. They saw their destiny as manifest—and glorious.

Settled habits dissolved, as a perceptive traveler from France, Alexis de Tocqueville, quickly noticed. A man might try ten occupations in turn, live in twenty different places. If he never reached European levels of mastery in a single skill, his experience sharpened his general intelligence and his political sense as well.

Trail-worn pioneers rest near Colorado Springs, Colorado, in the mid-19th century—a time of both expansion and inventiveness in America. Already, in 1836, as the Nation moved rapidly west, Congress had passed a law that for the first time protected inventors by giving them clear title to their creations. From then on, the United States Patent Office granted only patents that did not duplicate earlier ones.

DENVER PUBLIC LIBRARY, WESTERN HISTORY DEPARTMENT

Walter Hunt (1796-1859)
Resourceful Quaker

Even in the mosquito-ridden forests of the West, the pioneer kept his newspapers—brought by a postman on horseback—as handy as his ax. He kept up with civilization while he waited for it to catch up with him.

In this spread-eagle age, the United States acquired a new category of hero: the inventor. No longer a moonstruck figure of suspicion, the inventor was gaining dignity.

In a mood to do justice to such worthy citizens, Congress passed a new patent law in 1836. By the old procedure, the Patent Office had registered an invention on due notice and payment of a fee—and if two inventors claimed the same device, they had to go to court and fight things out. This not only clogged the courts, it also disheartened the inventors.

The Act of 1836 required proof that an invention was new, useful, and workable before any patent would be issued. It permitted extension of a patent for seven years beyond the usual fourteen, at the discretion of a special board.

Although much amended over the years, the 1836 statute marks the beginning of the modern American patent system.

With a new law to match the new spirit of the country, the number of patents issued began to rise—although some promising careers were promptly upset by the Panic of 1837 and the depression that followed.

Still, Americans had a notion that they could do anything if they tried hard enough and long enough. They could outshine anybody under the sun.

Proclaimed in bombastic orations and found true often enough, this swaggering faith carried men through the wilderness, through hard times, through downright calamities—sometimes to success. It inspired inventors to live up to the triumphs of Whitney and Fulton.

For example, it sustained Charles Goodyear in his struggles; as one modern scholar has observed with wry respect, his principal qualification was persistence.

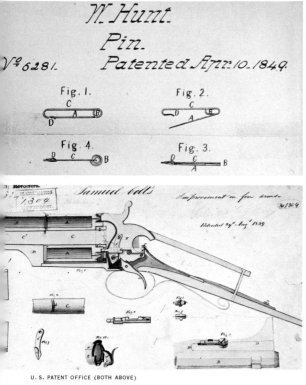

W. Hunt
Pin.
No 6281. Patented Apr. 10. 1849.

Fig. 1.
Fig. 2.
Fig. 4.
Fig. 3.

U. S. PATENT OFFICE (BOTH ABOVE)

First modern safety pin, patented by Hunt in 1849, evolved in three hours one afternoon as he sat twisting a piece of wire in his New York shop. Hunt sold all rights to the pin for $400. Samuel Colt's revolver (lower drawing), first patented in 1836, helped lead to mass-produced interchangeable parts. Below, "Men of Progress" appear in an imaginative painting.

From left: William Thomas Green Morton, demonstrated use of ether as an anesthetic; James Bogardus, helped make skyscrapers feasible by pioneering in the use of cast iron in construction; Samuel Colt; Cyrus Hall McCormick, reaper (model at his feet); Joseph Saxton, standardized Government weights and measures; Charles Goodyear (seated, arm on table), vulcanized rubber; Peter Cooper (behind Goodyear), built Tom Thumb, first American-made railway locomotive; Jordan Lawrence Mott (seated beside Goodyear), invented a coal stove; Joseph Henry (standing, left of column), induced electricity through electromagnetism; Eliphalet Nott (seated at base of column), invented a coal stove; John Ericsson (standing, right of column), developed the screw propeller, designed the Monitor; Frederick Ellsworth Sickels, mechanisms for steam engines; Samuel F. B. Morse, telegraph (hand on table near model of his telegraph recorder); Henry Burden, labor-saving devices, including a horseshoe-making machine used during the Civil War; Richard March Hoe, printing presses; Erastus Brigham Bigelow, power looms; Isaiah Jennings, dentistry tools; Thomas Blanchard, a tack-making machine and a lathe for turning gunstocks; Elias Howe, sewing machines.

SMITHSONIAN INSTITUTION (OPPOSITE, UPPER); CHRISTIAN SCHUSSELE, 1862, NATIONAL PORTRAIT GALLERY, SMITHSONIAN INSTITUTION (SHOWN IN PART)

CHARLES GOODYEAR
By Arthur P. Miller, Jr.

I F DESPAIR would ever conquer Charles Good-year, now was its chance. Hadn't the 38-year-old inventor spent more than four futile years in a vain attempt to solve the riddle of rubber—how to prevent the stuff from becoming sticky in summer, brittle in winter, and horrid-smelling in between?

Hadn't he and his assistant Nathaniel Hayward lost everything except their drafty old factory in Woburn, Massachusetts, when "impervious mailbags" on which they pinned their hopes sagged like their fortunes?

Wasn't the Nation still wracked from the Panic of 1837, rubber manufacture in disrepute, and Goodyear's long-suffering wife and four children penniless and hungry?

Friends advised him to try a sane business, to look after his family. But Charles Goodyear heard a different drummer.

Faith told him that God influences "some mind for every work." The Creator, he was convinced, wanted him to spend his life "trying to improve articles of necessity or convenience, for the use of man." Money-getting held no allure. Goodyear would gladly have given his inventions to the public rather than patent them—except that he needed funds for new experiments.

If he ever wavered, indecision evaporated that cold day in January 1839. After visiting his closed-down factory at Woburn, and stopping nearby to see his brother Nelson, Goodyear, ever restless, tried yet another experiment. He heated the same compound that had decomposed in the disastrous mailbags.

By mistake some fell onto a hot stovetop. Goodyear peered at the blob with unbelief. It did not melt as "gum elastic" always had, even in the lesser heat of an August day.

Instead, it solidified, "charred like leather," the inventor recalled. He realized immediately that somehow the compound had been transformed. Excitedly, he showed Nelson, who did not think this new phenomenon "worthy of notice." Charles was undeterred: "... if the pro-cess of charring could be stopped at the right point," he reasoned, "it might divest the gum of its native adhesiveness throughout...."

He dipped pieces of his compound into boiling sulphur and found that these, too, solidified. Along the edge of charred portions he discovered a line that seemed "perfectly cured."

To test its resistance to cold—that other buga-boo of rubber—the elated inventor heated a piece of sheet rubber, then nailed the swatch outside the kitchen door in the winter chill.

"In the morning," his daughter Ellen remembered, "he brought it in, holding it up exultingly. He had found it perfectly flexible...." After years of anguish—a quiet triumph.

Although no one recognized it that cold January day—except possibly the Yankee inventor—Charles Goodyear had discovered vulcanization. He had produced the first modern plastic, a new raw material that would serve where leather, wood, or metal would not.

Before Goodyear's discovery, rubber's bad qualities permitted few uses. Spanish invaders had found Indians in Central America using coagulated latex as balls in a sacred game; French savants had studied the new substance for waterproof qualities; someone had found that the gray gum rubbed out pencil marks on paper, and thus the word "rubber" was born.

By 1839 British manufacturers had learned a few other uses for uncured rubber. Charles Macintosh, a Glasgow chemist, patented in 1823 a fabric that included a thin layer of rubber. From this he fashioned raincoats that still bear his name. In the English climate such products satisfied their purchasers; in American winters they hardened like armor, in American summers they softened like taffy.

Thomas Hancock, a London coachmaker, cut garters, suspenders, and other items out of rubber. In 1844 Hancock would obtain a patent in England for vulcanization of rubber—after he examined samples produced by Goodyear.

Charles Goodyear, slight of build and short,

Charles Goodyear (1800-1860)
Vulcanizer of rubber

Portrait of Goodyear, painted at his direction on a thin sheet of rubber, recalls the inventor's long and persistent efforts to dramatize the versatility of his discovery. Father of today's vast rubber industry, he discovered vulcanization, the process that toughens rubber and rids it of stickiness, in January of 1839.

THE GOODYEAR TIRE & RUBBER COMPANY

to New Haven to become his father's partner.

In 1826 Charles, with his young wife Clarissa and baby daughter Ellen, returned to Philadelphia to open a hardware store and sell Amasa Goodyear's hay fork. The forks sold, but the business failed when Charles and Amasa extended credit too freely. With Yankee fortitude they refused to take the easy way out—bankruptcy. To pay their creditors they sold everything, and Charles finally delivered himself to debtor's prison as the law commanded.

In and out of jail as creditors made their claims, he turned to his workbench in free moments. Several inventions generated a meager living: a new patent button; an improved spring-steel fork; a spring-lever faucet; an air pump.

One day in 1834 Charles paid a visit to the New York City store of the original United States rubber manufacturer, the Roxbury India Rubber Company. His object: to peddle an improved valve for rubber life preservers.

The manager shook his head sadly. As a matter of fact, the firm would be lucky to stay in business. The "rubber fever" of the early 1830's had subsided as suddenly as it had begun. The public was fed up with products of the waterproof gum from Brazil. Shoes of raw rubber yellowed with age—they were dubbed "golden slippers." Invariably they belied advertisers' claims of "perfect" resistance to extremes of weather.

Thus challenge met Charles. He forsook the hardware business and "determined to make a profession of invention." In his early 30's, with a family to support and the odds heavily stacked against him, he determined to bend the obstinate gum-elastic to man's use. All his old fervor for rubber flooded back.

For a laboratory he used the kitchen of his cottage in Philadelphia. Mincing his raw rubber, he added turpentine as a solvent and kneaded the mass into a dough, adding one substance after another to rid the rubber of its natural stickiness. He mixed one smelly batch after another, spreading it on a marble slab with

frail, and sickly, seems an unlikely candidate for a giant of the modern rubber industry. He knew little chemistry, stood in awe of complicated machinery, and possessed small talent as a businessman. But curiosity he had, persistence in almost superhuman quantity, and unreasoning faith beyond the telling of it.

He grew up a Connecticut Yankee, eldest of six children of Amasa Goodyear, a New Haven merchant and sometime inventor. Amasa manufactured the first pearl buttons made in America and metal buttons that U. S. soldiers wore in the War of 1812; later he invented and produced a "Patented Spring Steel Hay and Manure Fork."

While still a schoolboy Charles grew fascinated with "India rubber," so called because it was believed to come from the West Indies. "It can be extended to eight times its ordinary length...," he marveled, "when it will again resume its original form.... Who can examine, and reflect upon this property of gum-elastic, without adoring the wisdom of the Creator?"

Finished with school at 17, Charles was apprenticed to a hardware firm in Philadelphia until ill health and overwork forced him back

In an experiment that failed, Goodyear tests rubber treated with magnesia in his kitchen laboratory. He finally

discovered vulcanization when a sulphur-impregnated batch accidentally fell onto a hot stove and hardened.

"Is Charles Goodyear the discoverer of this invention of vulcanized rubber?" asks Daniel Webster in Goodyear's behalf during the Great India Rubber Case of 1852. The eloquent Webster, then Secretary of State, came to Trenton, New Jersey, to protect his client against infringement by shoe manufacturer Horace H. Day. Webster won a decisive victory, but Goodyear, often involved in costly litigation and promotion, would profit little from his discovery.

ARTHUR LIDOV

Clarissa's rolling pin. He tried magnesia compounded in quicklime—and, it is said, salt, pepper, soup, witch hazel, ink. No luck.

Penniless again, the family sold its furniture, pledging Clarissa's homespun linen against a month's rent, and moved into lodgings while Goodyear went to New York City.

There fortune smiled wanly when he found that nitric acid kept rubber from getting so sticky. Exuberantly he made up rubberized cloth and rubberized paper. Hopefully he sent samples to President Andrew Jackson. A New York businessman advanced some $7,000 for Goodyear, who was no bookkeeper, to produce life preservers, hats, caps, and aprons, but the Panic of 1837 wiped out both backer and business.

Destitute, Charles and his family struggled along, living at an abandoned rubber factory on Staten Island and eating fish from the bay. He was so poor that one day he had to leave his umbrella as collateral for a ferry ride to Manhattan to see his pawnbroker.

Finally he raised a small loan and rented space in the Roxbury Company's factory, near Boston. Using his nitric acid process, he made attractive piano and table covers, carriage cloths, shoes, even maps from sheets of rubber.

In 1838 he joined Nathaniel Hayward to take over an idle rubber factory in nearby Woburn and produce life preservers, cushions, coats, capes, and tents. It was Hayward who introduced Goodyear to sulphur, dusting the odoriferous mineral over rubber and exposing the stuff to sunlight.

A formula combining Goodyear's acid treatment with Hayward's sulphur compound brought about the mailbag debacle. Only the surface of the rubber had been cured; the rest remained raw. Deteriorating items bounced back for refunds and this latest venture failed.

Now Goodyear's spirits hit bottom. Friends withdrew their sympathy. Financial aid dried up. Ridicule flayed the family. Hunger stalked it; neighbors gave the children milk.

It was at this low ebb that Charles Goodyear dropped his compound on the stove—and changed the course of his life.

If the years leading to the discovery were difficult, the next few years were wretched. Goodyear knew in his heart he had found the answer, but nobody would listen to him. In those years, as a rubber merchant testified under oath, "it was about death to a man's credit to be engaged in India Rubber business."

Still he persisted—alone. He knew that heat and sulphur together could miraculously change rubber. But how much heat? For how long? He roasted bits of rubber in hot sand, toasted them like marshmallows, steamed them over a teakettle, pressed them between hot rollers. When Clarissa took her bread from the oven, he thrust in chunks of evil-smelling gum. He hired a local bricklayer to build an oven—and paid the poor man partly with rubber products.

Poverty tested his tenacity. Writing in the third person, he confessed that "he collected and sold at auction the school books of his children, which brought him the trifling sum of five dollars; . . . it enabled him to proceed. At this step he did not hesitate."

He fished the streams to feed his family. When an infant son died, he could not pay for a coffin. (Of the twelve Goodyear children, six survived into adult life.)

At last he determined that steam or hot air applied for four to six hours at about 270° F. gave consistent results. At that heat the sulphur cured the rubber, increasing its strength and stiffness while preserving its flexibility.

Satisfied, he began producing shirred ribbons and other items, and a brother-in-law backed him. He slowly improved his process; slowly—very slowly—his goods began to find customers.

The days of destitution finally ended. Manufacturing rights could have made him a millionaire. Instead he set ridiculously low royalties on licenses he granted on his patents. He withdrew to dream up more new products so that these

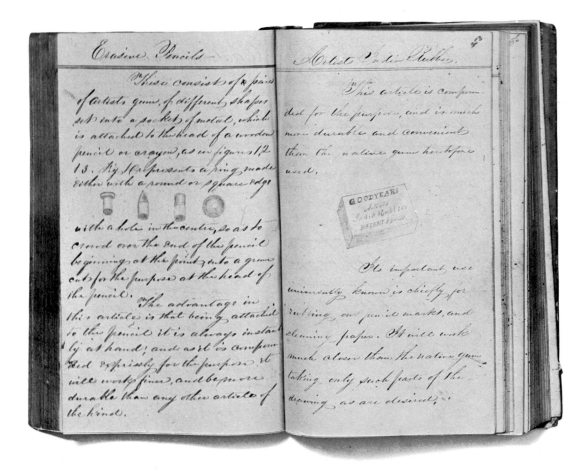

*Bound in the substance that absorbed his life,
Goodyear's* Gum-Elastic and Its Varieties, *published
in 1855, shows Amazon Indians tapping rubber trees.
The inventor's laboratory workbook at bottom,
opened to pages describing erasers, reveals that
Goodyear envisioned some highly imaginative uses
for his discovery: maps that would double as bed-
spreads or tablecloths, globes that would also be
"foot balls," tape measures, ponchos for milkmaids.*

VICTOR R. BOSWELL, JR., N.G.S. STAFF, THE GOODYEAR TIRE & RUBBER COMPANY

additional uses of rubber would not "escape the notice and attention of others, as the original discovery might have done . . . with less enthusiasm by the writer."

Against "patent pirates" Goodyear repeatedly had to prosecute infringement cases. In the Great India Rubber Case of 1852, his advocate was no less than the Secretary of State—Daniel Webster. The licensees paid Webster $15,000 to take leave from his job and plead for Goodyear. In a two-day speech at Trenton, New Jersey, Webster silenced Goodyear's chief tormentor— one Horace H. Day—and won a permanent injunction against him. The case made headlines —but the piracy of others continued.

Goodyear waited too long to apply for foreign patents. Thomas Hancock beat him to an English patent and he lost out on a technicality in France. It was even Hancock who popularized the word "vulcanization," derived from the Roman god of fire. The term held more magic than Goodyear's "fire-proof gum," and it caught on.

Today, more than a century later, chemists still imperfectly understand the molecular change that comes over rubber when it vulcanizes. They do know, as Goodyear knew, that under heat the compound grows stronger, more resistant to cold and heat, yet remains flexible.

Such characteristics have inspired a cornucopia of products that would delight the inventor. Many of them Goodyear foresaw—jotting one brainstorm after another into his notebook: rubber gloves, toys, conveyor belts, watertight seals, water-filled rubber pillows, balloons, printing rollers, rubber bands.

Goodyear also envisioned rubber banknotes, musical instruments, flags, jewelry, "imitation buffalo-robes," vanes or "sails" for windmills, and ship's sails, even complete ships. He occasionally wore a rubber hat, vest, or tie. The automobile tire did escape his imagination—only because the auto hadn't been invented.

He would find few surprises in a tire-manufacturing plant today. The formula remains much the same, although synthetic rubber has largely replaced natural rubber. Sulphur remains the chief vulcanizing agent. The huge mills that mangle raw rubber into a doughlike mass would doubtless remind him of endless hours spent kneading his obstinate "gum-elastic."

Charles Goodyear died in New York City at age 59. Ailing, he had stopped there on the way to New Haven to visit a daughter who lay critically ill. Long-suffering Clarissa, whose valiance matched Charles's single-mindedness, had died in England in 1853, her husband at her side.

In his autumn years, Goodyear had seen his discovery revitalize the American rubber industry. Factories began turning out items he had long before imagined. The Gold Rush of 1849 generated a demand for vulcanized boots—just the thing to withstand hours of panning for gold in a California stream.

But Goodyear never found the pot of gold at the end of his rainbow of products. Supremely unconcerned about money, he sank thousands into lawyers' fees in patent cases, splurged happily on promotion.

He spent $30,000 on a lavish entry for the great Crystal Palace Exhibition in London in 1851. Before the startled eyes of Europeans he spread his rubber marvels. Rubber globes filled with hydrogen floated high above his grand "Vulcanite Court." His display won a high honor, the Grand Council Medal.

For the Paris Exhibition of 1855, Goodyear created another extravagant showcase. When the Emperor Napoleon III gave him the cross of the Legion of Honor, it came to the cell at Clichy prison where the inventor languished temporarily for nonpayment of his latest debt.

Returning home, in poverty and failing health, he applied for—and won—an extension of his patent. In a long and eloquent report, the Commissioner of Patents cited "the generous spirit of the patent laws" and extolled Goodyear and inventors generally, "the true jewels of the nation to which they belong."

Automobile tires made by essentially the same process as that discovered by Goodyear—but now largely with synthetic rubber—move down a tire sorter (opposite) at the Goodyear Tire & Rubber Company in Akron, Ohio. The earthmover tire being lifted from its mold at left measures 9 feet 6 inches across and weighs 3,320 pounds. Of all the rubber consumed annually in the U. S. today, nearly two-thirds —3.85 billion pounds—goes into tire production.

Using a leg to brace himself, a tire maker of 1910 (below) pulls rubber-reinforced cotton fabric over a steel core before stretching on the tread. Goodyear died long before the development of the automobile and the founding of the company that bears his name.

ROBERT W. MADDEN (ABOVE); THE GOODYEAR TIRE & RUBBER COMPANY

"The Lightnin' Wire" — Telegraphy Conquers Time

5

SOME YEARS AGO, a *very* high government official sent a draft statement to the Smithsonian Institution for expert review. It included the sentence: "Every schoolboy knows Samuel F. B. Morse as the inventor of the electromagnetic telegraph."

The experts knew it wasn't so simple as that. They started composing accurate substitutes. Morse invented one of the first telegraphs? A successful telegraph? True, but not dramatic. Finally, triumphantly, they wrote: "Every schoolboy knows Samuel F. B. Morse as the inventor of the Morse telegraph." Both the official and his audience were happy, and the Smithsonian men could take special satisfaction in the episode because it was their Institution's first Secretary, Joseph Henry, who had laid the scientific basis for a critical part of Morse's work.

Surprisingly early, in 1753, an anonymous

Lines drawn taut, the gas-filled Enterprise *moors at the White House on June 18, 1861, after balloonist Thaddeus S. C. Lowe sent President Lincoln the first telegraph message from aloft. Joseph Henry (with papers under his arm), planner and first Secretary of the Smithsonian Institution, served as science adviser to Lincoln and encouraged him to exploit balloons as aerial observation posts during the Civil War.*

ARTHUR LIDOV

HENRY ULKE, 1874, NATIONAL COLLECTION
OF FINE ARTS, SMITHSONIAN INSTITUTION

Joseph Henry (1797-1878)
Foremost American scientist of his day

Briton had thought of a signal device with an "electrical machine a-going" and a wire for each letter of the alphabet, but no such telegraph was practical for decades.

A Frenchman named Claude Chappe introduced his visual telegraph, or semaphore, in 1792. Set on hilltop platforms about 15 miles apart, these signaling devices could send a message 150 miles in 15 minutes.

The first telegraph line on this principle in the United States was built in 1800, to transmit shipping news from Martha's Vineyard to Boston, and other lines soon followed.

In this setting Henry and Morse made America's first contribution to the telegraphic art.

JOSEPH HENRY
By Carroll W. Pursell, Jr.

A YOUNG HARVARD MAN looked up from his work one winter day in 1873. Into his room, cluttered with electrical apparatus, came a "dignified elderly gentleman" whose "remarkable philosophic countenance" reminded him of the European giants of science, Humboldt and Helmholtz. His description probably owed as much to awe as to close observation. In fact, Joseph Henry looked surprisingly youthful for his 76 years.

Perhaps it was the great physicist's pensive mood. After seeing the equipment and hearing about current research, Henry turned toward a table covered with telegraphic instruments. "If I had patented those devices," he remarked, "I should have reaped a large fortune."

It had been Joseph Henry's second near miss. In the late 1820's he had almost won credit for the discovery of an important law of nature; in the '30's he had nearly invented the telegraph.

Instead, he had to be content with a social invention perhaps more important than either — that of American science itself. Not that there was no science before him, or that he was responsible for all that flowered during his own

time. Henry was not the father — he was more the kind, wise, strict, and encouraging maiden aunt of the growing family of American science. He bore disappointment gracefully, for the most part, and with Christian patience.

Not surprisingly, some of his permanent traits were rooted in his early years. His grandparents had come from Scotland; his father was a day laborer, his mother a devout and strict Presbyterian. Joseph was born December 17, 1797, in Albany, New York; since his family was poor, he was sent for some years to live with a grandmother in the village of Galway, in the next county. There he attended school, and at ten became a clerk in a local store.

In his early teens he returned to Albany to live with his widowed mother. His very un-Presbyterian weakness for romantic novels degenerated into a love of the theater and he joined a local troupe of actors. From that dread career he was saved by the chance reading of a popular book, a Dr. Gregory's *Lectures on Experimental Philosophy, Astronomy, and Chemistry.*

In this book the inherent drama of nature shone through a style designed to whet the curiosity of the young: "You throw a stone, or shoot an arrow upwards into the air," wrote the author, an English clergyman. "Why does it not go forward in the line or direction that you give it?" Years later Henry wrote: "This book although by no means a profound work has under Providence exerted a remarkable influence on my life. . . . It opened to me a new world. . . ." He had promptly resolved to devote himself to knowledge, pausing only to present one final dramatic reading — a formal valedictory to his fellow thespians.

He threw himself into his studies, and in 1826 was offered the post of professor of mathematics

Covered with silk from his wife's petticoat, insulated copper wires coil around an experimental electromagnet built by Henry in 1831. The foot-high device could lift more than a ton suspended from the iron bar held fast across its poles. Power from a battery reached the coil through the curled copper leads.

and natural philosophy at the Albany Academy.

At his inaugural lecture, Henry declared his conviction that science should ameliorate man's condition, and discussed its role in "the mechanic arts." And almost immediately he began to experiment with electricity.

That science had not greatly advanced beyond Franklin's achievements, but in 1800 the Italian physicist Alessandro Volta had announced his new source of electricity, the voltaic pile: the first wet battery. For the first time scientists had a source of continuous current, and they soon began to discover new phenomena.

In 1820 the Danish experimenter H. C. Oersted reported his discovery that an electric current sets up a magnetic field around the metal that conducts it. Almost immediately, distinguished European investigators began to probe the new mysteries of electromagnetism.

Henry, too, the provincial American teacher, turned to the problem. Soon he discovered ways to increase the power and therefore the lift of such devices. In 1831 he published the different results he obtained by wrapping different numbers and lengths of wire around the soft iron core of the magnet. These he related to the size and type of battery best for a specific purpose.

By following his own rules he lifted 750 pounds with a magnet weighing 21 pounds — and as far as he knew, the previous record was held by an electromagnet in Philadelphia that weighed 52 pounds but could lift only 310.

News of probably the world's most powerful magnet reached the Penfield Iron Works at Crown Point, just north of Fort Ticonderoga. The firm ordered two of the professor's new magnets to separate iron from its ore. Soon after, the name of Crown Point was changed to Port Henry — an early honor for the man who struggled so hard to advance American science and bring it to bear on practical problems.

Next, Henry turned to the obvious question: How could one reverse the process and turn magnetism into electricity?

Henry's principle of electromagnetic induction — as shown in his experiment below — forms the basis for power production in generators, motors, and transformers. By raising and lowering the metal plates in the battery jar, the experimenter interrupts the current in coils around the electromagnet and induces electricity to flow in separate windings on the iron bar. The galvanometer registers the current.

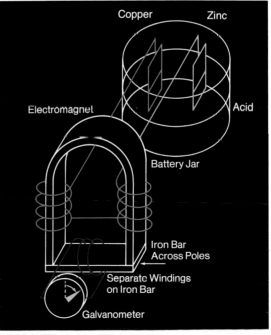

VICTOR R. BOSWELL, JR., N.G.S. STAFF, SMITHSONIAN INSTITUTION (UPPER); BEVERIDGE AND ASSOCIATES, INC.

Work on this was interrupted not only by his teaching duties and marriage to his cousin, Harriet Alexander, but also by a move, in November 1832, to the College of New Jersey at Princeton. In his years there he conducted experiments involving electromagnets, static electricity, the detection of electricity in clear skies — using kites — and thunderclouds.

Henry's research with thunderclouds and lightning was dramatically appropriate. His careful blend of experiment and theoretical reasoning on electrical phenomena set him apart as the first top-flight scientist produced in America since Franklin. It was fitting that years later his great English scientific rival, "prince of experimenters" Michael Faraday, proposed that the Royal Society of London award its coveted Copley Medal to Henry. Only one other American had received it: Benjamin Franklin.

As Henry pursued his leisurely investigation — paced by conflicting obligations and his own cautious nature — he was rudely jolted by news from London: Faraday's announcement to the Royal Society on November 24, 1831, of his own discovery of the current induced by magnetism.

One evening some 16 years later, a young scientist in Washington heard from the great Professor Henry's own lips how Faraday had anticipated him. Henry's apparatus, the young man noted, "had been prepared a year or more — consisting if I remember correctly — of a wire coil around the keeper [armature] of a steel magnet — but he was prevented from ... his experiment by interfering circumstances until he heard ... that Faraday had produced electricity from magnetism. He immediately concluded it must be the very thing he had been after and at once set up his apparatus — brought the keeper up to the magnet and the galvanometer needle at once whirled round and in a few minutes he obtained the spark."

The spark was a bonus. Not only had Henry confirmed Faraday's results, he had also discovered self-induction — the fact that a current passing through a wire induces a charge in the wire itself. When Henry published his experiments in July 1832, he established his own claim to the discovery of self-induction. Faraday repeated the discovery a few years later, apparently unaware of Henry's work.

Henry's daughter later commented that her father "often expressed his regret that he had neglected to publish his first results." No doubt that regret ran deep. In practical invention the prize goes to the winner in the form of a patent. In the more gentlemanly sport of science, priority of publication is equally important. Even to Henry, for whom science was a Christian calling, the laurel of priority was a prize worth seeking.

And there was no other. As one contemporary of Henry's put it: "No true man of science will ever disgrace himself by asking for a patent; and if he should, he might not know what to do with it any more than the man did who drew an elephant at a raffle. He cannot and will not leave his scientific pursuits to turn showman, mechanic, or merchant; and it is better ... that he should continue his favorite pursuits. ..."

This separation between the man of science and the practical inventor appears in another

Hydroelectric generators of Washington's Grand Coulee Dam (opposite) can each produce enough electricity for a city of 100,000. At left, water cascades 390 feet down the floodlit spillway. In the powerhouse of Colorado's Morrow Point Dam (below), water channeled by the huge pipe will turn the blades of a turbine connected to a generator on a floor above—inducing electricity through the same basic principle of electromagnetism pioneered by Joseph Henry almost a century and a half ago.

BUREAU OF RECLAMATION, DEPARTMENT OF THE INTERIOR

aspect of Henry's research. Shortly before his death, he recalled how he had applied his experiments in Albany "to the invention of the first electro-magnetic telegraph, in which signals were transmitted by exciting an electro-magnet at a distance, by which . . . dots might be made on paper, and bells were struck in succession, indicating letters of the alphabet." His telegraph line at Albany consisted of more than a mile of wire strung around a large room at the Academy. At Princeton he had returned to the problem. In his own words: "I think the first actual line of telegraph using the earth as a conductor [instead of wire, to complete a circuit] was made in . . . 1836. . . . signals were sent . . . from my house to my laboratory."

Late in life Henry remembered that a friend had urged him "to take out a patent . . . but this I declined on the ground that I did not then consider it compatible with the dignity of science. . . . In this perhaps," he concluded with ill-disguised regret, "I was too fastidious."

In both science and technology, the state of the art — the current level of both hardware and skill — often suggests further developments. And men with similar problems who turn to the same field of science for answers will find the same range of possible solutions before them. Thus "simultaneous" discoveries — and inventions — happen with fair frequency.

Morse and Henry clashed in later years over the amount and nature of help that the inventor received from the scientist. In part the quarrel arose from their different temperaments — Morse sanguine and combative, Henry cautious and retiring — but it also grew out of their self-imposed roles. The scientist was glad to be useful, but refused to turn mechanic or merchant. It was a matter of calling.

In 1846 Henry received a call of another sort. James Smithson had been a British savant, a Fellow of the Royal Society. Upon his death he left £105,000 — about $515,000 — to the United States to found an institution bearing his name,

for "the increase and diffusion of knowledge among men." Congress solemnly debated the propriety of accepting the gift and decided to do so. Then the question arose: how to spend it? Proposals ranged from a school of steam engineering to a library and lecture series.

At the request of the regents of the new Institution, Henry made a report that cut to the heart of the Nation's scientific needs. He proposed "to assist men of science in making original researches, to publish these in a series of volumes, and to give a copy of these to every first-class library on the face of the earth."

It was his greatest invention.

Henry's plan was approved and the regents chose him for the Institution's first Secretary, a post he held until his death in 1878. The Smithsonian's impact upon American science was enormous. Some critics begged — and some tried political influence — to absorb the endowment in schemes for popular entertainment and cultural uplift. But Henry kept the Institution on the high road of research support, and gained maximum advantage from his small funds.

At the Smithsonian, Henry interviewed young scientists and recommended them for jobs. By extensive correspondence he kept track of who was engaged in what research; he put investigators in contact with each other. He gently prodded the Federal Government to invest more generously in scientific activities. Such agencies as the Coast Survey and the Naval Observatory counted him as counselor and protector.

He took every opportunity to tap, for science, the new fortunes appearing in America. On a trip to San Francisco in 1871, he met the eccentric millionaire James Lick. Henry tactfully praised famous men, and pointed out that the patronage of science earned the gratitude of posterity. Lick responded by endowing the struggling California Academy of Natural Sciences, and gave $700,000 to build the world's largest telescope.

Nor did Henry neglect the faith he had professed in 1826, that science should "ameliorate

Sketch drawn by Henry in 1840 depicts one of his many experiments with electromagnetism. The coil, connected to a battery, magnetizes the horseshoe-shaped iron bar to create, at the instant of turning the battery current off and on, a ringing sound heard by the inventor through wires leading to his ears.

our present condition." The Smithsonian encouraged explorations of the West, and Henry took active interest in improving foghorns and lighthouses. During the Civil War he sponsored dozens of projects to bring science to the aid of the Union. When Thaddeus Lowe brought his balloon to Washington in 1861, he naturally sought Henry's aid in forcing military aeronautics on the attention of the harried War Department. In the midst of the war Henry helped found the National Academy of Sciences, and he served as its president from 1868 to 1878.

Henry was at his best in such activities. The young man who gave up acting had become a great impresario of science, commanding, directing, and encouraging an immense cast of individuals and institutions across a stage coextensive with the Nation itself.

Electromagnet powers a telegraph patented in 1845 by Englishmen Charles Wheatstone and William F. Cooke. A needle on the other side of the case, actuated by electrical impulses and watched by an operator, pointed to letters of the alphabet. In the U. S., the Morse system automatically recorded messages.

Once it was said of his famous contemporary P. T. Barnum that "people were as eager to see him as the great show he had created." Henry lacked that flair for self-publicity, but it made him no less indispensable.

Barnum, who knew a humbug when he saw one, reported sadly that "there is a scientific humbug just as large as any other." Henry saw many in his day—out-and-out frauds and sincere fanatics—and devoted a great deal of time trying to unmask their pretensions.

Encouraging real talent was a more congenial task. No episode better reveals this aspect of his service than his first meeting with Alexander Graham Bell. The young inventor arrived one frigid day at the door of the Smithsonian, anxious to interest the great Professor Henry in his plans. Despite a severe cold, the elderly Secretary was at his desk and took time to listen to the teacher from Boston. But for the encouragement he received from Henry, said Bell, "I should never have gone on with the telephone."

Henry's character matched the strength of his vocation. He lacked wit but not humor. He was sometimes laughably indecisive in small matters but his principles were rock-firm. His nationwide influence flowed from the universal respect he had fairly earned.

No American of the 19th century better served or represented the science of his time, and he molded science toward its modern shape.

Writing to one of his many friends in 1868, he mused tranquilly: "The individual is born and dies—the race remains. If the individual performs well his part...and leaves the world better than he found it, his mission is fulfilled."

Henry died knowing that he had fulfilled his mission well. As the Reverend James McCosh, President of Princeton, prayed at his memorial service, "We praise Thee because Thou didst put wisdom into his inward parts, and give understanding to his heart, so that he applied himself to seek out and to reach knowledge and the reason of things."

VICTOR R. BOSWELL, JR., N.G.S. STAFF, SMITHSONIAN INSTITUTION (BOTTOM); SMITHSONIAN INSTITUTION ARCHIVES

SAMUEL F. B. MORSE
By Carroll W. Pursell, Jr.

NINETEENTH-CENTURY WAGS were only half joking when they declared that invention was the last resort of all unsuccessful Americans. If this was so, Congress was the last resort of one unsuccessful inventor.

One day late in February 1843, Samuel Finley Breese Morse sat in the gallery of the House of Representatives, lost in discouragement, as Congressmen played the fool. At stake was an appropriation of $30,000 to build an experimental telegraph line which would prove the worth of his invention.

Cave Johnson from Tennessee rose to speak. Since the Congress was about to do something to advance the science of electromagnetism, he suggested wittily, he would like to see half the money go to a Mr. Fisk who was currently exhibiting the wonders of mesmerism—hypnotism or, as it was then called, "animal magnetism." He offered a formal amendment to that effect.

George S. Houston of Alabama then rose to urge support for the Millerites, a sect predicting the imminent second coming of Christ.

After this exercise in humor, the Chair insisted that the amendment must be voted upon, despite protests that it was all in fun. The motion was in order, ruled the Chair, and "it would require a scientific analysis to determine how far the magnetism of mesmerism was analogous to that to be employed in telegraphs."

An enterprising reporter found Morse sitting alone in the gallery, apparently unamused. With unprofessional understatement the reporter observed, "You are anxious." "I have reason to be," the inventor answered; "... I have spent seven years in perfecting this invention, and all that I had; if it succeeds, I am a made man; if it fails, I am ruined." Poignant enough for any age, for a man in his early fifties it bordered on tragedy.

It was an unlikely crisis for the life that had begun a half century before. On April 27, 1791, when his eldest son was born, the Reverend Jedidiah Morse was in his second year at the First Congregational Church of Charlestown, Massachusetts, where he preached orthodox Calvinism and practiced an equally orthodox Federalism. His battles against liberalism in religion and politics left him time to found the science of geography in America. His wife, too, was strong and intelligent; she bore eleven children and buried eight. Young Finley, as his friends called him, was served a nourishing intellectual diet along with the victuals at the family table.

After preparatory work at Phillips Academy, in Andover, he entered Yale in 1805, returned home briefly, then finally graduated in 1810. There he was stiffened by the daily sermons of President Timothy Dwight and exposed to some of the best science in the country.

Early in 1809 he wrote home that he found Professor Jeremiah Day's lectures on electricity "very interesting ... he has given us some very fine experiments, the whole class taking hold of hands, form the circuit of communication, and we all received the shock apparently at the same moment." The phrase "circuit of communication" stands out today, but Finley was no doubt innocent of its larger meaning.

For three years he heard the famous chemist Benjamin Silliman lecture on chemistry and galvanic electricity, and saw the new batteries invented by Alessandro Volta a decade earlier.

All in all, it was not a bad introduction to the subject. Almost a century later the Harvard physicist John Trowbridge guessed that "Morse evidently got all there was to be had at that time on the subject of electricity."

Nothing suggests that young Morse considered a career in science, not yet really a profession. Instead, he set his mind on becoming an artist. Whatever the innermost emotions of his father's heart, his parents wanted him to fulfill himself. When Washington Allston and Gilbert Stuart, two of the Nation's leading painters, praised his work, his father and mother agreed to send him abroad. In July 1811 Finley sailed with Allston to study under him in London.

War between Britain and America broke out a

Artist-inventor Samuel F. B. Morse, who painted the self-portrait at right, holds credit for devising America's first commercially successful electromagnetic telegraph. On May 24, 1844, with apparatus similar to the register, or receiver, he had patented four years earlier (below), Morse sent a message over an experimental telegraph line from Washington, D. C., to Baltimore, Maryland. He may have used the key at lower right to tap out the dot-dash signals.

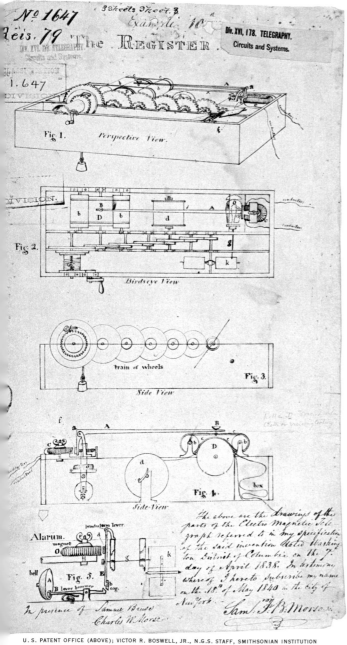

ADDISON GALLERY, PHILLIPS ACADEMY, ANDOVER, MASSACHUSETTS

Samuel F. B. Morse (1791-1872)

U. S. PATENT OFFICE (ABOVE); VICTOR R. BOSWELL, JR., N.G.S. STAFF, SMITHSONIAN INSTITUTION

This sentence was written from Washington by me at the Baltimore Terminus at 8 h 45 min A.M. on Friday May 24th 1844, being the first transmitted f W h a t h a t h G

"*What hath God wrought?*" reads the decoded paper-tape recording (above) of the historic message transmitted by Morse in 1844. Eight years earlier, in his New York City workshop (below), he assembled his first experimental electromagnetic telegraph system. The rectangular receiver (opposite) held a pencil

FROM SAMUEL I. PRIME, "THE LIFE OF SAMUEL F. B. MORSE, LL.D.," 1875 (ABOVE);
VICTOR R. BOSWELL, JR., N.G.S. STAFF, SMITHSONIAN INSTITUTION

in the A-shaped pendulum to make dot-dash code
marks on a paper strip pulled over a drum by the
clockwork. When cranked, the transmitter (bottom)
sent messages by opening and closing an electrical
circuit by means of the depressions and elevations
in the coded type moving under the rocker arm.

year after his arrival. In this innocent age he was not persecuted; he read the classics, socialized with fellow art students, visited with Benjamin West, and finally exhibited his work at the Royal Academy. "My passion for my art is so firmly rooted," he wrote his parents, "that I am confident no human power could destroy it. The more I study, the greater I think is its claim to the appellation of *divine. . . .*"

He returned to work in Boston in 1815, full of plans for opening a studio and becoming one of "those who shall revive the splendor of the fifteenth century." But in boisterous young America, busily shaping the 19th century to its own ends, art — like science — seemed slightly precious and aristocratic. Morse found that he had to become an itinerant painter of portraits. Even this proved less than profitable and, not for the last time in his career, the artist turned inventor.

With his brother Sidney he produced an improved pump for a fire engine. Eli Whitney thought it "would have a preference to those in common use," but financially it proved disappointing. In 1818 Finley wrote his parents: "The machine business . . . I am heartily sick of. It yields much vexation, labour, and expense, and no profit. Yet I will not abandon it. I will do as well as I can with it; but I will make it subservient to my painting, as I am sure of a support . . . if I pursue it diligently. . . ."

He needed financial security all the more because that year he married Lucretia Pickering Walker, of Concord. He found some demand for portraits in Charleston, South Carolina, but troubles accumulated. His father was driven from his pulpit by Unitarians and had to move to New Haven. Finley and Lucretia's second baby died, and his southern patronage began to decline. Coming north, he joined his young family at his parents' home. In 1823 he went alone to New York, already the most populous city in the United States and the center of artistic life in America. There he met with little more success than before. Then tragedy rained down upon him.

In 1825 his beloved Lucretia died, followed by his father in 1826 and his mother in 1828. In 1829 he departed for Europe to seek inner peace and renewal for his flagging career.

He remained three years, mostly in Paris and Italy, and then, in October 1832, sailed for home on the packet *Sully.* One night at dinner a talented but eccentric physician, Charles Thomas Jackson of Boston, brought up the subject of electricity. Morse asked whether an electrical impulse could travel over a long line; Jackson assured him this was possible.

Morse — as he remembered later — exclaimed: "If this be so, and the presence of electricity can be made visible in any desired part of the circuit, I see no reason why intelligence might not be instantaneously transmitted by electricity to any distance." After dinner he promptly started sketching a crude telegraph system.

Arriving in New York, however, he had other pressing concerns. He taught painting and sculpture at the University of the City of New York. He became involved in the anti-foreign, anti-Catholic "Know-Nothing" movement spreading over the country in the wake of growing immigration. In 1836 he ran for mayor on the "Native-American" ticket, but drew fewer than 2,000 votes. A year later, Congress rejected his bid to help decorate the rotunda of the Capitol. It was the culmination of years of disappointing struggle for recognition as an artist. Once again he resolved to concentrate on invention.

Already many miles of semaphore telegraph were in operation in the United States, and a proposal that Congress finance a semaphore system connecting New York and New Orleans attracted Morse's attention in 1837. He knew this system had one great drawback: It was totally ineffective at night and in poor weather. His own method, using electricity, avoided this defect and he warmly recommended it to the Congress.

Morse's system had evolved steadily since that night aboard the packet *Sully.* Before the

Fixing a bracket to a tree, a lineman in 1859 strings the "lightnin' wire," so-called because it carried electricity. By 1861, when lines extended across the country, the telegraph had replaced the Pony Express.
FROM TALIAFERRO P. SHAFFNER, "THE TELEGRAPH MANUAL," 1859

end of 1836 he had completed a crude device.

His original sender consisted of a kind of printer's composing stick, a tray which contained saw-toothed pieces of type arranged to make and break an electric circuit in a properly coded order. The receiver consisted of a frame holding a pencil that made marks on a moving strip of paper when agitated by a current passing through an electromagnet. Despite its crudeness it was simple, and it worked.

Morse had the personal qualities to succeed as inventor of the telegraph: intelligence, persistence, a willingness to learn. What he lacked was equally important: knowledge of recent scientific developments, adequate capital, mechanical ability, political influence. Like all successful inventors, he exploited his strengths and worked to repair his weaknesses.

His first lack was met by Professor Leonard D. Gale, who taught chemistry at the city university. Gale suggested that Morse improve both his battery and his electromagnet by following the suggestions of Joseph Henry. Together they incorporated Henry's discoveries—and stepped up the distance they could send messages from 50 feet to 10 miles.

At this point young Alfred Vail, a recent graduate of the university, joined the enterprise. Vail was a fine mechanic, with access to his father's iron and brass works in New Jersey. For a quarter-interest in the invention, he agreed to help develop it, and produced improved, simpler, sturdier instruments. Gale also became a partner, and Morse pressed forward with his patent application.

He took his apparatus to Washington in February 1838, and it won the admiration of F. O. J. Smith, a slightly disreputable Congressman from Maine. Smith offered himself as lawyer and publicity agent in exchange for a partnership. He promptly used his position in the House to recommend a grant of $30,000 to Morse.

Now Morse set off for Europe to gain foreign patents and backers. He failed. Rival inventors acknowledged the superiority of his device; but in England the system of William F. Cooke and Charles Wheatstone, though quite different, was awarded priority. In Paris he caused a sensation in scientific circles, where his only competitor for attention was Louis Daguerre with his new process of photography. Knowledge of that process was the most Morse had to show for his trip when he returned to America in 1839.

He got through the next few years of frustration partly by teaching the art of daguerreotypes to students in New York—among them a young dry-goods clerk named Mathew Brady.

Late in 1842 Morse made one more trip to Washington to seek money from Congress. He wrote to his brother Sidney of his desire to cease being a burden to his family and "to have the means of helping others." His old friend, the sculptor Horatio Greenough, noted: "Poor Morse is here with his beautiful, his magic telegraph. . . . He goes regularly to the House."

Finally both House and Senate passed his bill,

LIBRARY OF CONGRESS, BRADY COLLECTION

Accompanying the Union Army into battle, civilian
linemen of the U. S. Military Telegraph Construction
Corps string wire near Brandy Station, Virginia, in
1864. Wagons carrying reels of wire drove from
pole to pole as the men worked. Union operators at
left rest outside the telegraphers' tent after the
Battle of Gettysburg. The first great military body to
demonstrate the advantages of the field telegraph,
the Union Army put up some 15,000 miles of wire during the war—compared to 1,000 for the Confederacy.

To link the Old World and the New, the converted British man-of-war Agamemnon *stows nearly 1,300 miles of transatlantic telegraph line in a cable-laying project conceived in 1854 by American business magnate Cyrus W. Field. In 1866, after more than 12 years beset with heartbreaking defeats, the British steamer* Great Eastern *completed the task; in her test room (opposite, below) scientists and engineers wait tensely for the reply to a signal sent to Ireland.*

and Morse had $30,000 to build an experimental line. On May 24, 1844, the wire from the railroad depot in Baltimore to the Supreme Court Chamber in the Capitol was ready. Morse sat quietly in Washington and began sending with his new transmitting key: "dot dash dash—dot dot dot dot—dot dash—dash—...."

Forty-one miles away Vail decoded the message: "What hath God wrought?" The telegraph was a proven success.

On May 29 word flashed by wire from the Democratic convention in Baltimore that James K. Polk had been nominated for the Presidency.

At first the telegraph created more drama than business. People flocked to see the magic key at work, but could think of no messages to send. To fill the time, operators finally decided to promote chess matches between Baltimore and Washington players, but such frivolity caused stirrings in the religious community. In a typical bit of American logic, it was decided to report Congressional doings instead.

As early as 1848, however, every state east of the Mississippi, except Florida, was served by telegraph; by the end of the Civil War more than 200,000 miles of line were in use for business communications and personal messages as well as news of battles, politics, and sports results.

The business end of the enterprise proved more difficult than the technological. Once Gale had told Morse of Joseph Henry's improvements in magnets and suggestions for batteries, science disappeared from telegraphy, not to appear again until the laying of the Atlantic cable. Such important components as insulators for the wire were improved on a strictly cut-and-try basis.

Samuel F. B. Morse was just 53 when the Congress voted money for his line. He had already had a long life for the period, and one darkened by personal tragedy, frustrated artistic ambition, and inventive promise unfulfilled. A decade later he was financially secure for the first time in his life. At last he could provide for his family, in-

cluding the children of a second marriage, as he wished; but tranquillity still escaped him.

Patent litigation, quarrels with his partners, pressure from competitors, public clashes with Professors Henry and Jackson, and the growing national controversy over slavery all served to sour his increasing years.

No dispute was so sordid or unnecessary as his campaign against Joseph Henry, who had always come to his aid when Morse had sought scientific advice or public support. Then, in 1849, during a patent suit, Henry filed a deposition denying any knowledge "that Mr. Morse made a single original discovery" in electromagnetism which applied to the telegraph. "I have always considered," he concluded, "his merit to consist in ... the invention of a particular instrument and process for telegraphic purposes."

This statement was accurate but ungenerous. Of course, the invention rather than the discovery was in question. As someone later said of Morse, he had had "the faculty of seeing the value of corner lots when other men were lost in contemplation of the surrounding scenery."

Always passionate in controversy, Morse proceeded to belittle Henry's scientific contributions and to undermine in public what he privately called Henry's "jackdaw dreams" of glory. It was cruel but characteristic of Morse. The breach was never healed.

In the years of old age, before his death in 1872, Morse began to enjoy the success so long withheld from him. He had outlived most of his antagonists. His peace efforts during the Civil War, marked with racial prejudice and pro-Southern sympathy, were forgotten in the bustling expansion of the Grant era.

At last for a brief hour he became what he had always wanted to be—simply the "means of helping others." He was, after all, one of the great inventors of all time; the man who, in the phrase of William Cullen Bryant, had "annihilated both space and time in the transmission of intelligence."

NEW YORK PUBLIC LIBRARY (ABOVE); METROPOLITAN MUSEUM OF ART, GIFT OF CYRUS W. FIELD, 1892

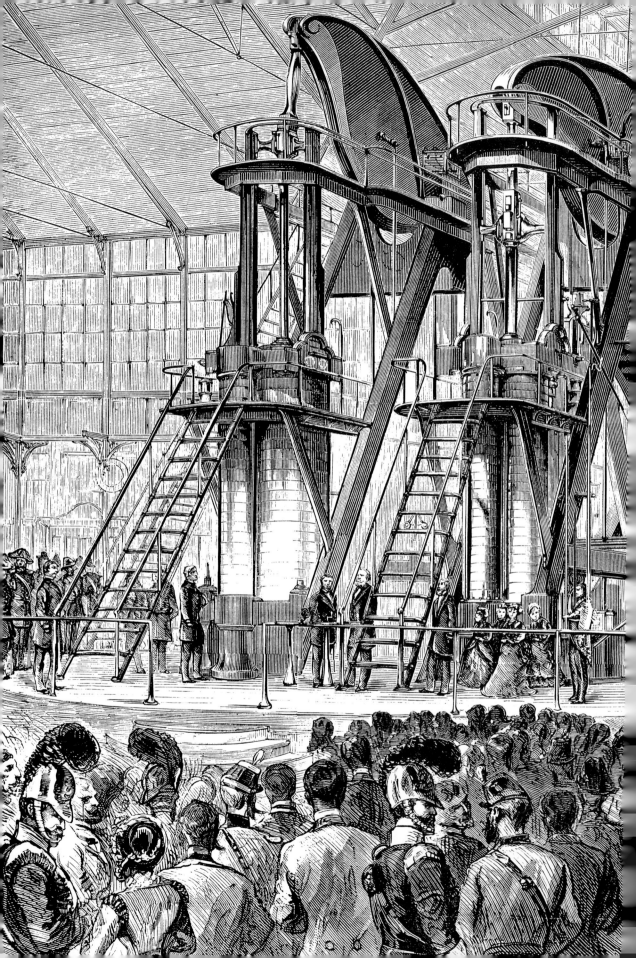

A Centennial Gallery: Inventions and National Life

6

"THE POOREST LITTLE SHANTY in the town had its penny flag hung out," an eyewitness remembered, when the patriots of 1876 opened their celebration of the Nation's first 100 years: the Centennial Exhibition in Philadelphia.

Bunting and banners, lank from hours of rain, quickly dried in the breeze as the sun came out on the morning of May 10, and thousands of visitors fought their way through a classic traffic jam to the grounds of Fairmount Park. Vendors cried their wares—"Ice lemonade!" "Hot roast potatoes!" "Get your official guidebook, right here!" Reporters scribbled their notes, ladies drabbled their ruffles in the shrinking puddles, gentlemen kept their good temper in eddies of jostling and pushing.

At a special platform before Memorial Hall, citizens and foreign dignitaries—"heavy swells," in the slang of the day—struggled to their seats

President Ulysses S. Grant and white-bearded Emperor Dom Pedro II of Brazil (at foot of center stair) start the great Corliss steam engine, opening the U. S. Centennial Exhibition of 1876 in Philadelphia. Built in Providence, Rhode Island, the engine furnished power for displays in Machinery Hall, "and without a moment's derangement," recalled an observer, "drove thirteen acres of machinery."

HARPER'S WEEKLY, 1876, LIBRARY OF CONGRESS

LIBRARY OF CONGRESS (TOP); HISTORICAL SOCIETY OF PENNSYLVANIA

Flags snap in the breeze as crowds arrive at the Main Building, reportedly the largest such structure in the world. Eight million of the country's forty million people celebrated their Nation's first century with a Centennial tour. "Why, along at first when I was a beginnin' my tour through the United States," declared a character in a book about the Exhibition, "I would be fearfully surprised at the awfully grand and beautiful things; but before noon I got so that I wasn't surprised at nothin'."

Columbia proudly gestures toward her accomplishments — railroads, steamboats, democratic government, and the Centennial — in an allegorical engraving that appeared on the cover of Frank Leslie's Illustrated Historical Register of the Centennial Exposition.

America's new breed of heroes — inventors — offer their talents to Liberty personified on a stock certificate issued by the Centennial (left). U. S. citizens subscribed directly more than $2,250,000, and indirectly, through Congress, some $2,000,000 more.

U. S. DEPARTMENT OF AGRICULTURE

Onetime slave George Washington Carver, who transformed the South's agriculture, works at Tuskegee, Alabama. His 47 years of research there, beginning in 1896, persuaded cotton farmers to plant other crops and won him international renown. He developed more than 300 products from peanuts alone.

for the opening ceremonies. Hearty cheers greeted the ebullient Emperor of Brazil, for Dom Pedro II, simply dressed and "as devoid of ornament as a freight car" according to one reporter, appealed to contradictory elements of American taste: fascination with the alien grandeur of a crown, and touchy insistence on the national variety of down-to-earth democracy.

In this complicated mood the people of the United States welcomed the world to their anniversary celebration—and waited for the world's acclaim. "Inviting all the other principalities of the globe to...a competitive display," said a typical spokesman, the Republic could show

"the greatest progress that had ever been made in the world's history in the same length of time —an advancement without a parallel...."

At the proper moment President Ulysses S. Grant—about the only public man in the country not given to the florid oratory of the day—made a terse speech. He summed up the century past: the wilderness cleared, the works of civilization established. Now America, he said, would show how far she could rival "older and more advanced nations" in learning and art: "Whilst proud of what we have done, we regret that we have not done more."

In due course, inside Machinery Hall, the President and the Emperor set the magnificent Corliss steam engine into action. Around them, with a clatter and droning, 13 acres of machinery went to work: spinning cotton, shaping wood, printing newspapers and wallpaper, sewing and polishing shoes.

One nonchalant engineer tended the Corliss as it supplied the power for 8,000 machines, and novelist William Dean Howells decided that "in these things of iron and steel...the national genius most freely speaks...."

It spoke an American idiom, for belt-and-pulley systems—not the heavy shafts and toothed gear wheels of Europe—transmitted that power in American factories. Oliver Evans had run his flour mill this way; his countrymen had adopted the method, with its loose-jointed efficiency, by the 1840's. It might look slapdash, but it was economical to maintain. It served the manufacture of interchangeable parts undertaken by Whitney and others. Now, when foreigners said "American system," they meant mass production by techniques these men had pioneered.

Thanks to the inventive genius of the Yankee, as home-grown commentators remarked over and over, American style in farm equipment and laborer's tools meant light weight and tough construction; in machine tools, novelty and progress; in sewing machines, pre-eminence. (Already the Patent Office had issued fully two

BAKER LIBRARY, HARVARD UNIVERSITY (TOP); BEINECKE RARE BOOK AND MANUSCRIPT LIBRARY, YALE UNIVERSITY

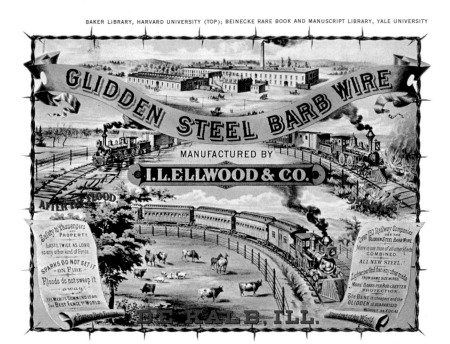

Colorful poster of 1877 advertises an American invention that inspired feuds between cattlemen and farmers, but conquered the West. The Glidden company dominated the market by buying out competitors.

Josiah Willard Gibbs, Yale's professor of mathematical physics from 1871 until his death in 1903, exemplifies the rare American theoretician. Few U. S. scientists knew enough math to understand his work.

thousand patents for sewing-machine inventions and improvements and attachments.)

Visitors stared as workers from the American Watch Company demonstrated division of labor, repeating selected stages from the thousand operations of producing a cheap machine-made watch. "With remunerative employments for ladies so limited," an open-minded young man was glad to see women exercising their "fineness of touch" in this industry.

Delighted by the abundance of consumer goods and proud of business enterprise, few if any who paused at such exhibits considered them a threat to democracy. Alexis de Tocqueville had asked, years earlier: What happens to the worker who does nothing for 20 years but make heads for pins? "The art advances, the artisan recedes," he answered. The workman becomes a new kind of serf, the master a new aristocrat, almost an emperor. Friends of democracy, Tocqueville warned, "should keep their eyes anxiously fixed" on industrial change.

For the six months of the Exhibition, some eight million Americans—from a population of forty million—feasted their eyes on an exhausting collection of marvels: a Gorham tea service worth $3,500; glass cylinders of black loam from Iowa; white furs from Norway; photographs of

NATIONAL GALLERY OF ART, INDEX OF AMERICAN DESIGN (TOP); HAYNES FOUNDATION, BOZEMAN, MONTANA (OPPOSITE); SOUTHERN PACIFIC COMPANY

Lace antimacassars protect mohair seats in this painting combining details of railway palace cars of the late 1800's. One line's advertisement, in the March 1896 NATIONAL GEOGRAPHIC, proclaimed: "LUXURY SPEED and SAFETY are COMBINED to conduce to the PLEASURE and COMFORT of its PATRONS."

Officers and workmen of the Central Pacific and Union Pacific railroads celebrate the driving of the golden spike at Promontory Point, Utah, on May 10, 1869. By 1897 four more railroads spanned the continent, bringing settlers to the West and providing them with easy access to the markets of the East. A daring feat of engineering, the wooden trestle spanning Marent Gulch in Dakota Territory supports an engine and the all-important pay car.

Derricks sprout near Oil Creek in Pennsylvania in 1884 on fields supplying oil for lubricants and lamps. The automobile had created a greater market for petroleum by 1921, when Texas gusher Robinson Number One (upper) came in near Houston. Canadian-born Elijah McCoy obtained 52 patents, most of them for devices to lubricate running machinery.

COURTESY MILTON MELTZER

Elijah McCoy (1843-1929)
He brought automation to oiling

everyday life in Australia; 15-foot cornstalks from Kansas; Mexican onyx; Liberian ivory; a $1,157 cashmere shawl from India; a Nevada quartz mill; shimmering glass from Bohemia; Italian statues; the world's largest soda fountain; carpets and laces from Belgium; superb Chinese carvings; Daniel Webster's old plow, the "constitutional soil expounder"; porcelains and bronzes displayed with Japanese restraint.

Critics dismissed much of the current art as rubbish, but in the Women's Pavilion crowds gathered before a painting of Truth unveiling Falsehood while Ignorance grovels in the dust. They eagerly inspected a high-relief sculpture of a dreaming princess; a Mrs. Brooks of Arkansas had carved "Sleeping Iolanthe" in good local butter and set it on ice in a tin frame.

At the Exhibition or at home, citizens groped for a new sense of national purpose, worthy of a new century. True, the Union was secure, at the cost of half a million dead; despite a lingering depression, its industrial output surpassed five billion dollars. The postwar decade seemed both energetic and degraded. Mark Twain, who had given the Gilded Age its name, jeered that America had "legislatures that bring higher prices than any in the world." Corruption tainted government until, shouted one orator, "the man in the moon has to hold his nose as he passes over the earth."

In the South, night riders of the Ku Klux Klan cast a white shadow. Reconstruction, a tangle of hope and boodle, anger and botch, had almost outworn the national patience. Elsewhere, politicos in need of votes waved the bloody shirt of the Union dead. Sweetly or bluntly, women asked why, if ignorant freedmen could vote, ladies could not. Businessmen fretted over the economy and talked nervously of growing trade unions and the prospect of socialism.

Students from Boston and Harvard Universities, who published the *Centennial Eagle* in 12 numbers on their own initiative, made a full report on one of the most popular exhibits at the

LIBRARY OF CONGRESS

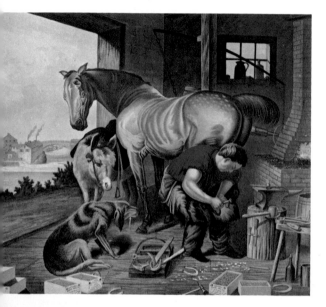

Exhibition: an early New England home, labeled "Ye Olden Time" and built of logs at a cost of $1,600. (Nobody pointed out that such houses were clapboard.) Citing such "ancient curiosities" as a Revere mold for pewter spoons, a rusty sword carried at the Battle of Concord, and a cross-stitch sampler, they explained that everything at "this primitive hut" represented "the oldest and oddest style imaginable."

In much the same tone of wonder and condescension, the young gentlemen commented on the "miserably carved idols" from the British Gold Coast colony and the relics of the Pharaohs. Perhaps in fact no country had ever lived through so long a century and emerged so changed.

Assessing the Centennial as it drew toward its wistful close, reflective men drew different conclusions for the Nation. A bishop compared American work to European, and decided: "we are too much in a hurry to do things handsome." The Chief of the Bureau of Awards, which bestowed medals and prizes, said that the crowds —far from well-informed, generally, but alert and shrewd—"afforded perpetual wonder" and "were the most notable thing exhibited." A politician, who might have pointed out that the Corliss engine dominated popular attention, proclaimed that steam had liberated more people than the Declaration of Independence.

Far from Philadelphia, a new and legendary America was taking shape, out on the cattlemen's frontier. Billy the Kid was rustling stock; Frank and Jesse James escaped a Minnesota posse; playing poker in a Deadwood saloon, Wild Bill Hickok was shot in the back; George A. Custer and his command died at the Little Bighorn; working cowboys took bawling longhorns up the dusty trails.

Prospects of new fortunes, new reforms, new inventions remained bright. Americans could say, with the editors of the *Centennial Eagle:* "let us read on every side the noble enterprise of the worthiest of mankind, and glory in the excellent characteristics of our age!"

Enraptured by their "Light Running Domestic
Sewing Machine," newlyweds ignore reception
guests. Industry in the late 1800's increasingly turned
its attention to the home, producing labor-saving
appliances—novelties that soon became necessities.
A newfangled "Home Washing Machine & Wringer"
(opposite) liberated women from backbreaking wash-
boards. Men, too, benefited from the flood of manufactures. The village blacksmith drives a mass-
produced "Ausable horse nail"; years before, he
would have used nails hand-cut from sheet iron.

LIBRARY OF CONGRESS

Arthur Lidov

Magic to Order: Electricity for America

7

"A PHILOSOPHICAL TOY," Joseph Henry sensibly called the little electromagnetic motor he invented in 1831. As late as 1876 he dismissed it as impractical compared with steam—its metals and battery acids cost far too much.

But other inventors—among them a Vermont blacksmith, an English clergyman, and a German savant working in Russia—had been improving such devices. From Henry's "toy" and from the primitive generator devised by Michael Faraday, the modern electric motor and dynamo took shape over half a century.

In the 1880's English, American, and Hungarian inventors developed the transformer. Now electricians could step up voltage for transmission and lower it to safer levels for use. The civilized world was entering the Electrical Age—and in the United States the heirs of Franklin were putting his legacy to work.

Using a mouthpiece connected to a human eardrum and middle-ear bones (in plaster mount before him), Alexander Graham Bell makes speech visible for deaf students. The figurative cross section above Bell shows how his phonautograph works. Vocal sounds vibrate the eardrum and bones; an attached straw traces the pattern onto a moving strip of smoked glass. The realization that electricity might carry such undulatory patterns led to his invention of the telephone.

ARTHUR LIDOV

COURTESY EDINBURGH SOCIETY

Alexander Graham Bell (1847-1922) as a youth of 20
He brought speech to the deaf—and to machines

ALEXANDER GRAHAM BELL
By Robert V. Bruce

"FIVE MINUTES' CONVERSATION is about as much as thirty pages of letter paper...," wrote an Edinburgh scientist to an American friend in April 1871. And he added fretfully, "All the boasted civilization of the 19th century has not been able to give us ... an equivalent for a chat over a quiet pipe." Professor P. G. Tait spoke for his times. He felt a need for conversation between continents—and impatiently complained at the slowness of its coming.

While his letter traveled slowly to America, the man suited above all others to meet that need reached the place best suited to his work. The man was from Tait's own Edinburgh by way of London and Canada, a Scotsman with an American temperament, a tall, dark-haired, dark-eyed, incorrigibly dramatic young teacher of speech named Alexander Graham Bell. And the place was Boston, the leading center of American science and technology.

Both the theory and the technology needed for long-distance transmission of speech had been at hand since the triumphs of Joseph Henry, Michael Faraday, and Samuel Morse, well before Bell's birth on March 3, 1847. But something had evidently been missing in the men or the times or both.

Judging from Bell himself, the inventor needed enthusiasm, ambition, intelligence, and imagination—but more specially, a fine sense of pitch, unusually keen hearing, thorough knowledge of the mechanics of speech and hearing, and expertness in acoustics, telegraphy, and the piano. His place of work would have to provide leading scientists, skilled technicians, enterprising capitalists, and an academic community. When Bell went to Boston in April 1871, his interests and gifts found a responsive setting.

Bell's paternal grandfather had turned from cobbling shoes in St. Andrews to acting in comedies in Edinburgh and finally to teaching speech in London. Bell's father carried the family name higher still, not only by writing and teaching but also by inventing what he called

"Visible Speech"—a system of written phonetic symbols capable of distinguishing any sound the human vocal organs could produce.

Both the dominating personality and the professional eminence of the father may explain in part the strong drive of young Alexander Graham Bell for recognition in his own right—what his younger brother teasingly called "his wish to do something great."

As a boy, Aleck—second of three sons—played at science and invention. He dreamed of fame as a concert pianist; his playing was excellent, but his growing interest in the sciences lured him away from the necessary hours of practice. After he graduated from the Royal High School in Edinburgh, his formal education was spotty.

Nevertheless, tension between him and his father revealed the intensity of his urge for eminence and figured in his first step toward telephone fame, in 1865. He and his father had disagreed about the natural pitch of the mouth cavity as it formed various vowels. He thought of sounding tuning forks before his open mouth while shaping it for a particular vowel sound. The fork that sounded loudest, by the principle of resonance or sympathetic vibration, would be the one with the natural pitch of the mouth cavity. This experiment proved, surprisingly, that each vowel is a compound of two pitches—and that both the Bells had been right.

Bell's father received the news with gratifying

Bell taps a message into the palm of 21-year-old Helen Keller during her 1901 visit to his summer home near Baddeck, Nova Scotia. He had helped direct the education of the deaf-and-blind girl, who later became famous as a lecturer and author. Bell's father (center) and John Hitz, Superintendent of the Volta Bureau for the Deaf in Washington, D. C., wait to "speak" to Miss Keller; Bessie Safford, Bell's secretary, takes notes. On Baddeck Bay in 1914 Bell listens attentively to his deaf wife Mabel, whom he tutored in speech and lip reading when she was a girl.

respect and admiration. Aleck invested in more tuning forks and whanged away exuberantly. He wrote up his discovery for his father's friend, the distinguished phonetician Alexander Ellis. Ellis dashed Bell's hopes of early fame with the revelation that the German physicist Helmholtz had reported the same findings, but young Bell remained elated because he had matched, independently, one of the greatest scientists living.

Ellis told Bell something else. Helmholtz had devised a means of keeping several tuning forks in continuous vibration at once with electromagnets in a battery-powered circuit. (In America, this gadget was called a rheotome.) Misconstruing the German account casually translated for him by Ellis, Aleck Bell got the notion that Helmholtz's device actually received and transmitted sound. And he jumped to the conclusion that speech also could somehow be telegraphed —not that he saw himself as the man to do it.

© THE BELL FAMILY (ABOVE); GILBERT H. GROSVENOR

Some years passed while Aleck taught speech at Bath and then became his father's assistant in London. He learned more about telegraphy, and a less extravagant vision came to the forefront of his mind: that of a multiple telegraph, one which could transmit several messages concurrently over a single wire. Such a device would make its inventor rich.

Thin walls separated his father's London parlor from the adjoining houses, each with its own piano. Bell was aware that a chord struck on one piano would sound on another if the felt dampers were lifted. Here again was an instance of sympathetic vibration. Reasoning from this, he saw that the combined frequencies of a chord could be transmitted electrically over a single wire and re-created by an array of tuning forks. Each fork could receive its own message at its own frequency—thus, the multiple, or "harmonic," telegraph.

But tragedy struck the family before Aleck Bell could put his notion to the test. First his younger brother died of tuberculosis in 1867, then his older brother in 1870.

ARTHUR LIDOV (ABOVE); © THE BELL FAMILY

"I hear, I hear!" exclaims Emperor Dom Pedro II of Brazil as Bell demonstrates his *"speaking telegraph"* at the Philadelphia Centennial on June 25, 1876. Exhibit judges crowd close; Bell holds his apparatus at right. His notebook (right) records the first telephone message, to his assistant, on March 10, 1876, at Boston: *"Mr. Watson—Come here—I want to see you."*

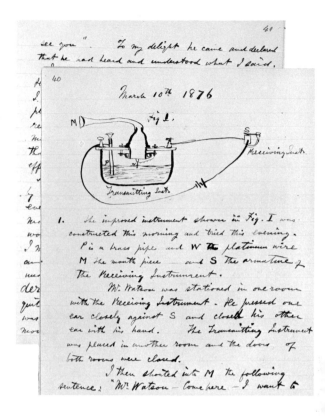

His father had lately toured the United States and Canada, lecturing on Visible Speech. There he had found a climate free from the smog and drizzle of London and Edinburgh. He had also found honor beyond that accorded him in his own country, and a spirit of enterprise that excited sympathetic vibrations in his own temperament. As a friend put it, he "caught the 'go-ahead' infection the short time he was there."

In 1870 the elder Bell and his wife decided to migrate to the New World and they pressured Aleck into coming, fearing that he might otherwise succumb to his brothers' disease.

The Bells settled in Brantford, Ontario, that August. But the restless Aleck seized a chance to teach deaf children in Boston, using Visible Speech as he had done successfully in London. With irresistible charm and enthusiasm, he helped convince Bostonians that those born deaf could learn to speak. The plight of children walled from the world by deafness touched Bell's heart—in his old age he would still give his profession as "teacher of the deaf."

Presently he won appointment at Boston University as a professor of vocal physiology. This gave him access to an academic community that included the scientists at the Massachusetts Institute of Technology. Evening lectures on physics at MIT, and newspaper reports on the success of a "duplex telegraph" invented by Joseph B. Stearns of Boston, probably had combined by October 1872 to set Bell once again on the trail of a harmonic telegraph.

Bell's idea was simple. Translating it into practice turned out to be something else again. Slowly, painfully, secretively—though with some help from academic friends—Bell labored to build a dependable rheotome, to work out circuitry, to develop efficient receivers. He switched from tuning forks to simple strips of steel that he called "reeds." Meanwhile he kept on teaching.

In the fall of 1873 he acquired a new pupil, Mabel Hubbard, a spirited and charming girl almost 16, who had totally lost her hearing from scarlet fever at the age of 5 and had almost lost the ability to speak as well. Accustomed as she said "to the dainty neatness of Harvard students," she wondered at first if Bell with his careless dress was a true gentleman, but she found him a superb teacher.

Her father, Gardiner Greene Hubbard, found him worthy of help. A successful lawyer and financier, he had promoted state support for aid to the deaf and also, vainly, a quasi-governmental telegraph system to compete with Western Union. In October 1874 Bell told Hubbard about his harmonic-telegraph scheme. At once Hubbard joined with Thomas Sanders, father of another of Bell's pupils, to finance experiments and patent expenses. Thus Bell could draw more freely on Charles Williams's shop in Boston for custom work in electrical apparatus. There he met a bright young workman named Thomas Watson who became his right-hand man.

"No finer influence than Graham Bell ever came into my life," wrote Watson half a century later. "His punctilious courtesy to every one was a revelation." The young electrician carefully copied Bell's table manners, read the scientific texts that Bell introduced him to, diligently practiced correct speech as Bell explained it, and did

Bell escorts Orville (holding door) and Wilbur Wright to their car outside the Smithsonian Institution in Washington, D. C., on February 10, 1910. The brothers had just received the Institution's Langley Medal for their achievements in aviation. Smithsonian Secretary Charles D. Walcott (far left) accompanies them.

his best to improve the devices Bell wanted. "His head seemed to be a teeming beehive...," Watson remembered. "A dozen young and energetic workmen would have been needed to mechanize all his buzzing ideas." Watson never forgot the clarity with which Bell explained his idea for a *speaking* telegraph.

At Brantford during his summer vacation in 1874, Bell daydreamed about Mabel Hubbard, with whom he had begun to fall in love, about the multiple telegraph, and about the phonautograph, a device that translated sound waves into an undulating curve on smoked glass. He hoped to use this to help deaf pupils "see" the sounds they uttered. Suddenly his jumbled thoughts fused into an epochal perception.

A rheotome did not vary the strength of its battery current, except by interrupting it at fixed intervals. So the receiver produced only a simple monotone. But if the strength of an *unbroken* current could be made "undulatory" like a sound wave, it would make the power of an electromagnet vary accordingly. And the magnet's varying pull, exerted on a diaphragm, would produce a sound with all the subtle modulations and overtones required for speech.

But how to create such a "sound-shaped" current? Bell saw an answer in theory, from Henry's work: A diaphragm armature vibrated by the voice would *induce* such a current in another electromagnet.

Thus a sound could create a current and a current could create a sound. But Bell assumed, without trying it, that such a current would be too feeble to work. And that fall Hubbard—with the authority of a sponsor and patron—brushed aside the whole notion and urged him to pursue the multiple telegraph.

So Bell and Watson labored on the telegraph that winter, spurred by indications that a Chicago inventor named Elisha Gray was hot on the trail of the same thing. In February 1875 Bell—who had applied for American citizenship—hurried to the Patent Office in Washington, only to find that Gray had filed a harmonic-telegraph patent two days ahead of him. In the end, Edison's quadruplex took the field anyway, but while in Washington, Bell told Joseph Henry about his telephone experiments.

Henry reacted with vision that belied his age. Doubtless remembering his own "too fastidious" reluctance to patent the telegraph, he urged Bell not to make the same mistake. When Bell protested that he lacked the technical knowledge of electricity to perfect his invention, Henry replied simply, "Get it!"

So that spring Bell kept thinking about a telephone. He thought of another way to make his undulatory current: using sound waves to vary the *resistance* in a battery-powered circuit. But the means he tried—varying the tension in a piano wire—did not work.

Then, up in the attic of the Williams shop on the hot afternoon of June 2, 1875, struggling with a multiple-telegraph circuit, Watson plucked free a receiver reed from its electromagnet. Beyond a partition, Bell saw another reed vibrating and heard it twang. At once the significance broke upon his mind.

He charged wildly in to the astonished Watson and confirmed a surmise—the mere motion of Watson's reed had generated an effective current by induction alone. Listening repeatedly to the sound that resulted, he became convinced that it was in the same realm of complexity as human speech. His "magneto-telephone" had not yet conveyed words—but it had a voice.

That summer Bell told the Hubbards and his parents his wish to marry Mabel, and confessed his love to her. Whirls of emotion—arising from family hesitations, the uncertainties of a girl of 17, and her father's insistence on the harmonic telegraph—strained him almost to despair.

Not until Thanksgiving Day did the young couple become engaged, and not until January did Bell complete his telephone patent application. Hubbard filed this on February 14, 1876.

By one of the most remarkable coincidences

NATIONAL AIR AND SPACE MUSEUM, SMITHSONIAN INSTITUTION

in the history of invention, Elisha Gray filed a sketch of a similar instrument as a caveat—or statement of an untested idea for an invention—only a few hours later. But Bell's patent was duly issued on March 7, 1876—Patent No. 174,465, often called the most valuable single patent in history. Bell, Hubbard, and Sanders set up their own company and got operations under way.

Ironically, the best-known episode in the Bell story was relatively unimportant. On March 10, his patent in hand, he returned to experiments with his variable-resistance concept. He used a platinum needle sticking down from the transmitter diaphragm into a weak acid solution, and found that as the needle dipped and rose it varied the strength of a current passing through needle and water to an electromagnetic receiver.

Bell noted the result: "I then shouted into M [the mouthpiece] . . . 'Mr. Watson—Come here—I want to see you.' To my delight he came and declared that he had heard and understood what I said." This now-famous first intelligible message by telephone did not even get into print until 1882; and Watson's recollection of years later that Bell called him after spilling some acid on his trousers was probably transferred from some other occasion.

Anyway, history would have been no different if Watson had been out for a coffee break at that moment; for on March 27 Bell succeeded in transmitting speech with his magneto-telephone also. This time it was his father who heard it; and appropriately enough the first word was "Papa." Thenceforward father and son got along splendidly; and this transmitter was the basis of the first commercial telephones.

Loving a dramatic effect, Bell in later years made much of another famous episode: demonstration of his telephone at the Centennial Exhibition in Philadelphia in June 1876, before the judges and Dom Pedro II, Emperor of Brazil. He told his grandchildren the judges on that hot Sunday were ready to quit before reaching the telephone, but Dom Pedro (who had met Bell in Boston) insisted on speaking into the device, to an aide in another room, and picturesquely exclaimed, "My God, it speaks Portuguese!"

Dom Pedro indeed took delight in the invention. But in fact the judges were duty-bound to examine the exhibit, and Bell had discussed it beforehand with the member of the party who really mattered, the great English scientist Sir William Thomson (later Lord Kelvin). Thomson's tributes to Bell made the English as well as the

1893 May 25th Thursday — at Wash. D.C. 235

Mr. Ellis & I will touch off each lamp by means of iron rods. Each rod dipped into sulphuric acid. On signal will simultaneously touch off stbd. & port number with sulphuric acid — and fly

"1, 2, 3"

and again

now we will try it — and note results.

© THE BELL FAMILY (ABOVE); GILBERT H. GROSVENOR

Tackling the problems of flight, Bell designs kites in his laboratory near Baddeck in the early 1900's. Their triangular-cell structure proved light, strong, and stable, but created too much drag for use as flying machines. Some ten years earlier, Bell had investigated rocket propulsion. The cartoon-like sketch he drew on May 25, 1893, shows Bell and an assistant igniting cups of fuel attached to a rotor blade. As it whirls, the men take refuge behind trees; it later exploded. At right, White Wing, christened for her white cotton covering, scatters spectators as she rises into the air near Hammondsport, New York, on May 18, 1908. Built by Bell and fellow members of the Aerial Experiment Association, she flew 1,017 feet in 19 seconds three days later. Also concerned with the down-to-earth problems of local farmers, Bell (left) feeds a ewe from his own flock. By selective breeding, he developed a strain that bore twins or triplets more than half the time.

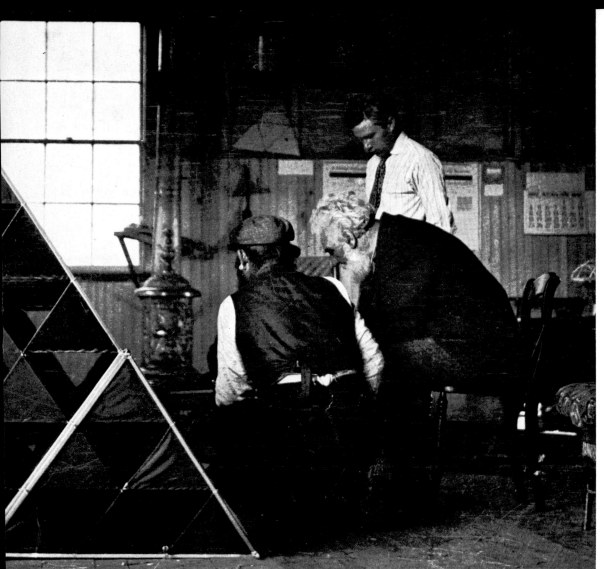

JOHN A. D. McCURDY (ABOVE); © THE BELL FAMILY

Rising on ladderlike hydrofoils, HD-4 *skims over Baddeck Bay. Bell's assistant, Casey Baldwin, based his design for the craft on the inventor's research and ideas. On September 9, 1919, twin aircraft propellers astern sped her at 70.86 miles an hour—setting a world speed record unbroken for ten years.*

Americans take notice—although when Professor Tait in Edinburgh heard about Bell's invention, he said, "It is all humbug, for such a discovery is physically impossible."

Bell and Watson resumed work on the telephone, and Bell relished the job of publicizing it in lecture tours. After he married Mabel Hubbard in July 1877, however, the couple sailed for England and Scotland, where they stayed more than a year. His technical work in telephony ceased entirely after the early eighties; he cheerfully admitted that keeping up with its technology did not appeal to him. In technology, as in people, Bell preferred the very young.

Thanks to Hubbard, Sanders, and others, Bell after 1876 had ahead of him 45 years of time and money to do precisely what he chose. He used those years well, sleeping late but working later —till 3 a.m. He organized and led a crusade for teaching speech to the deaf; Helen Keller dedicated her autobiography to him. "He is never quite so happy," she noted, "as when he has a little deaf child in his arms."

As a six-year-old, not only deaf but blind, she had felt his tenderness when her father took her to him for advice on her education. As a college student she defined his charm: "He makes you feel that if you only had a little more time, you, too, might be an inventor...."

He financed the early stages of Albert Michelson's famous measurements of the speed of light, work which figured in Einstein's theory. He also financed the first dozen years of *Science,* now the principal journal of the American scientific community as a whole. He played an important role in the development of the National Geographic Society.

"Wherever you may find the inventor," he said in 1891, "you may give him wealth or you may take from him all that he has; and he will go on inventing." In 1880 he invented the "photophone," which transmitted speech along a beam of light. It found only limited commercial use, although—with characteristic indifference to received opinion—he always insisted it was his greatest invention.

Bell's mind was fertile in notions. Some were magnificently absurd, like the phonograph-alarm for pedestrians; swung in a circle, it would yell "Help!" in case of felonious assault. Some, brilliant in their simplicity, were developed independently by other men years later, like the vacuum-jacket respirator—an "iron lung"—first successfully tested by Bell in 1881, and the literal "sounding" of ocean depth by timing underwater echoes, a suggestion of 1879.

As he neared 60, he thought of a system of construction based on tetrahedral frames, mathematically unsurpassable in their combination of lightness, strength, and simplicity. He patented this system in 1907, but did not live to see it developed in the famous geodesic domes of inventor-and-architect R. Buckminster Fuller.

His well-publicized flight experiments helped prepare the public mind for the triumph of the Wright brothers, and his Aerial Experiment Association of 1907-09 brought forth not only the first airplane to fly in the British Empire, but also a group of pilots that included Glenn Curtiss.

In his old age Bell wrote a series of articles for deaf children on "Simple Experiments." "Few men," noted the editor, "possess the gift of seeing things from the viewpoint of a child so clearly as Alexander Graham Bell." In this truth lay his weaknesses: extravagant enthusiasms, fits of impatience, an absorption in whatever currently occupied him that sometimes seemed inconsiderate. In it also lay his essential strengths: freshness of vision, contagion of spirit, scrupulous concern for truth, unquenchable hopefulness and thirst for knowledge, and an inexhaustible capacity for wonder.

These were the traits of an American. And he proclaimed himself proudly an American by choice, rather than by accident of birth. After his death on August 2, 1922, at his beloved summer home in Nova Scotia, his grave was marked with the inscription "Died a Citizen of the U.S.A."

GILBERT H. GROSVENOR (TOP); © THE BELL FAMILY

Experimenting with a forerunner of the iron lung in September 1892, Bell records that it "seemed to succeed perfectly." A metal vacuum jacket fit snugly around the chest of one of Bell's assistants, who "stated that he made no effort to breathe." As the bellows pumped air into and out of the jacket, air pressure within the jacket alternately squeezed and released his chest. A scrap of paper, fluttering at his mouth, proved that air moved in and out of his lungs. Bell thought the device might save premature babies and revive victims of near-drowning.

THOMAS ALVA EDISON
By Robert Evett

Delicate filament of carbonized cotton sewing thread glows within a replica of Thomas Alva Edison's first successful incandescent lamp. It introduced electricity to the American home. At left, the container for an Edison cylinder, forerunner of the disc phonograph record, bears a portrait of the man whose name became a worldwide synonym for "inventor."

THOMAS ALVA EDISON enrolled at birth in what the inspirational literature of his time described as the school of hard knocks. He triumphed over handicap, hardship, and adversity to become rich, famous, and legendary. Almost 30 years after Edison's death, L. Sprague de Camp wrote: "With 1,093 patents to his name, he was the most productive inventor in the history of the United States, and possibly the most productive in the history of the human race."

Edison came as close as anyone can to embodying the American Dream of Horatio Alger. True, he smoked cigars, chewed tobacco, and talked dirty. Worse, he scandalized the pious with his impersonation of the village cracker-barrel atheist. Nonetheless, he lent himself to the sloganeers of his period, and added a couple of slogans of his own. "There is no substitute for hard work," he said. "Genius is 99 percent perspiration and 1 percent inspiration."

Born in 1847, youngest and most mischievous of his family, he grew up in an atmosphere of small-town harshness. Nancy Edison—who "was the making of me," said her son—reserved a birch switch for him behind the old Seth Thomas clock and beat him with it until the bark eventually wore off. When at age 6 he set a fire in the barn "just to see what it would do," his father, Sam Edison, flogged him publicly in the town square; and he kept on whipping him long

after the boy began paying his own way. Did Edison's energy, self-discipline, and drive stem, at least in part, from the punishment he took as a child? Perhaps—he rarely spoke of it.

At 12, a grammar-school dropout with just three months of formal education behind him, he started working as a candy butcher on a train, hawking newspapers and cheap luxuries on the run between Port Huron, Michigan, where he lived, and Detroit. When he could he took refuge in the Detroit Free Library, where he read everything that fell under his hand. In later years he remembered especially *The Anatomy of Melancholy* and—in translation—*Les Misérables,* which he vastly enjoyed. Not so Newton's *Principia.* "It gave me a distaste for mathematics from which I have never recovered," he said.

Before his railroading days, Edison had played in the cellar of his house with whatever scientific equipment he could get. Sam came down from time to time to give him a licking, declaring that his chemicals would blow them all to kingdom come. "Al"—as his earliest friends called him—just might have burned the house down. With time, he wisely moved a good bit of his laboratory to the baggage car of the Grand Trunk Railway, where a tolerant management let him experiment until a fire gave them second thoughts.

Apparently as a result of scarlet fever and untreated ear infections, the boy developed a progressive deafness. In one of the most poignant sentences he ever wrote, he said, "I haven't heard a bird sing since I was twelve years old."

Birds or not, he heard some things. He invented the phonograph and enjoyed listening to it—sometimes biting the horn of the speaker so the vibrations of sound would be carried by the bones of his head—and, until he died at the age of 84, could converse with his second wife. When he had almost lost the whole world of sound, people would tap words in Morse code onto his knee or into the palm of his hand—at 11 he had started practicing the code on his homemade telegraph set.

WILLIAM L. ALLEN, N.G.S. STAFF, HENRY FORD MUSEUM, DEARBORN, MICHIGAN (BELOW); COURTESY H. BART COX

Edison takes his phonograph to the Nation's Capital: After demonstrating his wonder at the Smithsonian Institution, at the National Academy of Sciences, and to Members of Congress, Edison visits the White House. The machine so fascinated President Rutherford B. Hayes, his wife, and several guests that Edison's visit did not end until 3:30 a.m.

Stylus of an 1878 phonograph indents sound patterns into tinfoil wrapped around its turning cylinder. To reproduce the sounds, the needle retraces its path and transmits the vibrations to a diaphragm.

Edison as a 14-year-old newsboy in 1861

In the winter of 1863, when he was 16, Edison got his first grown-up job. The Civil War had created a demand for proficient telegraphers, and he spent several years knocking around the country, working at outposts of Western Union. Because he could never abide living by the rules, he never kept a job for long. Often he was down to his last dime.

He amused himself by fiddling with the telegraphic apparatus, at that time still desperately in need of refinement. After some very near misses, at 21 he invented a recording stockticker, immensely useful to speculators in gold and securities of the postwar boom. He wrote to his father that he had heard "mother is not very well and that you have to work very hard. I guess you had better take it easy after this. Don't do any hard work and get mother anything she desires. You can draw on me for money."

He produced the stock-tickers himself, in a rented workshop in Newark. For manufacturing fine equipment in an age when electrical engineering hardly existed as a profession, Edison needed, and got, superior craftsmen. A 21-year-old American mechanic, John Ott, who took a job with the company, remarked later that Edison was "as dirty as any of the other workmen, and not much better dressed than a tramp." Ott spent most of his life working for Edison, and participated in all of the great discoveries. So did a British mechanic, Charles Batchelor, and a young Swiss clockmaker, John Kruesi. These three shared the excitement, if not the glory, with Edison in the busy years just ahead.

Morse's telegraph could send only one signal at a time, and in one direction, and this struck everybody as eminently reasonable. However, since the 1850's there had been some experiments at sending more than one message each way. One of Edison's very first projected inventions was a duplex telegraph that, if luck had allowed him to finish it sooner, would have emerged as a triumph. Another inventor, Joseph B. Stearns, beat him to it.

ARTHUR LIDOV (ABOVE); WILLIAM L. ALLEN, N.G.S. STAFF, HENRY FORD MUSEUM, DEARBORN, MICHIGAN (OPPOSITE, UPPER); EDISON NATIONAL HISTORIC SITE, WEST ORANGE, NEW JERSEY

Edison's former assistant for photography (below, left), on a trip to Europe, meets a pioneer in motion study in Paris about 1898. Dickson's work under Edison produced the kinetoscope—a battery-powered film viewer, forerunner of movie theaters. Films of the 1890's featured a man sneezing and two employees dancing as Dickson recorded a violin solo.

Sometimes with money and sometimes without, Edison battered away at the notion of a multiplex telegraph. Finally, in 1874, he achieved a machine that could send and receive four messages at once, two going and two coming. His biographer Matthew Josephson has described this as "the masterwork of his youth."

In the spring of 1876, Edison—prosperous and crowding 30—bought land in Menlo Park, a New Jersey hamlet about 25 miles southwest of New York City. His inventions had brought him fame and money, but also trouble: the lawsuits that plagued him throughout his life. He needed a good place to work, and he established what Josephson calls "the first industrial research laboratory in America, or in the world, and in itself one of the most remarkable of Edison's many inventions."

Edison put his own savings into the project, and one job applicant reported that he warned, "Well, we don't pay anything, and we work all the time." He ran Menlo Park as firmly as an abbot in the dark ages would have ruled a community of monks. Hardly anybody got a night's rest. As an itinerant telegraph operator, Edison had learned to take naps wherever he was—under a table or on a roll-top desk—and he expected his associates to adapt to his schedule. "My children grew up without knowing their father," said John Ott. "When I did get home at night, which was seldom, they were in bed."

In the summer of 1876, Alexander Graham Bell enjoyed great public success with the demonstrations of his telephone. The moguls of Western Union, who had declined to support Bell, engaged Edison to invent a better version.

What Edison did create was a transmitter superior to Bell's liquid or magneto versions—it used a small disk of carbon to vary the electrical current. Another inventor, Emile Berliner, had entered a caveat in the Patent Office that anticipated some of the details in the transmitter Edison tried to patent two weeks later. Although the resulting patent tangle was not resolved

HISTORY OF PHOTOGRAPHY COLLECTION, SMITHSONIAN INSTITUTION

W. K. L. Dickson (1860-1935), Etienne Jules Marey (1830-1904) Practical developers of a flickering fantasyland

for 15 years, Western Union agreed to leave the telephone field to the Bell firm.

Edison said bluntly that Western Union had pirated Bell's receiver and the Bell company had pirated Edison's transmitter. Years later he told a subordinate: "Everybody steals in commerce and industry. I've stolen a lot myself. But I *know how* to steal. They don't—and that's all that's the matter with them."

His most original invention came out of experiments with an embossing recorder for telegraphy, in 1877, when he stumbled onto the principles of the phonograph. Simply, he found that if

WIDE WORLD (TOP); EDISON NATIONAL HISTORIC SITE, WEST ORANGE, NEW JERSEY

Able to nap anywhere, Edison snoozes during a camping trip near Hagerstown, Maryland, in 1921. President Warren G. Harding (in shirtsleeves) and Industrialist Harvey S. Firestone read the news. On another trip three years earlier, to the Great Smoky Mountains, Edison and Harvey Firestone, Jr., stand on the hub of a waterwheel. A friend, Professor R. J. H. DeLoach, sits before them; bearded naturalist John Burroughs and Henry Ford perch atop the wheel. Harvey Firestone, Sr., stands at right.

he took a vibrating diaphragm from an experimental telephone, attached a blunt pinpoint to it, ran a piece of waxed paper under it, and shouted into the transmitter, the pinpoint would leave indentations in the wax. It would follow that if he ran the paper through another machine with a similar stylus and diaphragm, he would get the sound back.

On July 18, 1877, he bellowed *"Halloo!"* into the most primitive phonograph imaginable and, with Charles Batchelor, heard some vaguely recognizable human response. Four months later he paid John Kruesi $18 to make a machine out of brass and iron with a revolving cylinder turned by a hand crank. They put a piece of tinfoil around the cylinder, and Edison, in his high-pitched voice, recited one full stanza of "Mary Had a Little Lamb" into one of the diaphragms. Many a brave cigar was lost that day in bets over whether the thing would work or not. It did.

Edison reaped a golden harvest of publicity, which he loved, and took the machine to Washington, where he showed it off to Joseph Henry, Representatives and Senators, and President Rutherford B. Hayes. It earned a lot of money as a sideshow attraction that recited music-hall jokes; but when, after a couple of years, its vogue passed, Edison turned to other projects and let it languish for a decade, until his most active creative phase had passed.

In the summer of 1878, Edison, for the first time in his life, took a vacation. He went to Wyoming to observe a total eclipse of the sun. On that trip, conversations with a traveling companion fired him with the idea of creating a workable electric light. Sir Humphry Davy had demonstrated an arc light in London 70 years earlier, and since then many people had realized the potentialities of electricity for illumination.

A British scientist, Joseph W. Swan, was experimenting with incandescent lamps before 1850, when Edison was still a toddler. Both inventors found that carbon, excited by electrical energy, makes light. Edison differed from others

EDISON NATIONAL HISTORIC SITE, WEST ORANGE, NEW JERSEY

working on the incandescent lamp in that, as a man of action, he wanted to create a marketable product, and to create it in a hurry. In the beginning, he could only have imagined dimly the host of auxiliary inventions needed to make the lamp a commercial success.

After working himself and his men up to 20 hours a day for weeks on end, Edison made an incandescent lamp that yielded a reddish glow and burned for 40 hours. That granddaddy of electric lights began to shine on October 19, 1879. Like all light bulbs, it burned out, but only after Edison had deliberately increased the voltage to see how much it would support.

To get his lamp out of the laboratory and into the world, he had already embarked on that imperative series of inventions required for generating electric power and delivering it to the consumer. He completed the basic work on this system, one of his greatest accomplishments, in only three years of trial and error.

In big cities, where almost everybody had gaslight already, it was no great problem to thread electric wiring along the pipes. Edison had merely to improve on existing dynamos to have a cheap, reliable source of power. But he had to invent many things from scratch: devices for sealing the bulbs, screw-in sockets, light switches, safety devices, meters, and a host of other contrivances.

Working like Santa Claus and his elves, Edison and his men had a pilot system ready in 15 months. "After this," he said, "we will make electric light so cheap that only the rich will be able to burn candles."

Conceding that it would take time "to keep the bugs out" of his lighting system, he made it work in New York. He bought some rundown real estate on Pearl Street in lower Manhattan, and built his first power station. It opened on September 4, 1882, with one dynamo servicing only 85 subscribers. By 1884, New York had 508 consumers and a constantly growing demand. Power stations were going up in other cities

and the Electrical Age, though not yet flourishing, had come into existence.

A British commentator points out that Edison "straddled two American worlds—the technicalities of its 19th century industry and the jungle of its financial barons. . . ." In this jungle during the 1880's, Edison ranked at least as a baronet. A more fastidious age might blench at the sharp practice of the electrical industry and Edison's lack of interest in the people who worked for him. But in the late 19th century, few expected anything else.

Edison took with him to New York his first wife, Mary Stilwell. She had married him when she was only 16, and spent the best years of her short life at Menlo Park with her three children, taking tea with the ladies—the wives of the few married men in the Edison plant—and eating herself into obesity. Not long after the wedding he had jotted in a notebook, "My wife Popsy-Wopsy can't invent." In New York he encouraged her to play the elegant lady; she returned in summer to Menlo Park where, in 1884, she contracted typhoid fever and died. The grieving Edison, at 37, had completed more than half of his life's work in invention and had 47 years to go.

As a young man he had gone beardless—at a time when men cultivated full beards. Later he prided himself on being just a country boy who had made good, and also on his eccentricities of speech and dress. As he grew older, his once-charming Huck Finn manners took on overtones of grossness. He liked to play practical jokes, and to spit on the floor. Once, offered a spittoon, he pointed out that if you spit directly on the floor you never miss.

VICTOR R. BOSWELL, JR., N.G.S. STAFF

Eccentric genius Charles Steinmetz explains an insulator for high-tension wires to Edison during the inventor's 1922 visit to General Electric's laboratory and plant at Schenectady, New York—the former Edison Machine Works. Despite his jibes at theoretical scientists, Edison liked Steinmetz—a mathematician and physicist—because "he never mentioned mathematics when he talked to me."

Worse, his anti-intellectualism became a passion. He always liked to twit mathematicians and theoretical scientists, but the time came when, if he didn't like the form progress was taking, he simply turned his back on it.

The classic case was the great AC-DC controversy. By the mid-1880's, with the development of a practical transformer, George Westinghouse and others demonstrated that alternating current has this advantage over direct current: More power can be transmitted over longer distances at higher voltages. Edison had vast sums tied up in DC equipment, and when New York State decided to dispatch malefactors by electrocution, Edison urged them to use, by all means, the "Westinghouse current" rather than his own, implying that it was lethal and Edison's current safe. Westinghouse was furious at this campaign of hokum, and Edison, as he later realized himself, was blundering.

In 1886 Edison found himself another bride, Miss Mina Miller of Akron, Ohio. She displayed all the refinement a bourgeois lady could wish— while her husband, one of the most famous men in the world, willfully created the image of frontier coarseness itself. *Collier's Weekly* ran a story called "She Married the Most Difficult Husband in America." Perhaps she did. But, in the teeth of her husband's opposition, she gave her children a formal education appropriate to their condition in life.

In later years Edison embarked on projects that consumed far more time and money than those of his youth and rewarded him with far less uniform success. He lost a fortune and ruined some New Jersey countryside tearing down a mountain in the hope of extracting iron from low-grade ore. He spent years on a storage battery for an electric car; but the gas buggies from Detroit took the field. He patented a cheap, prefabricated cast-poured concrete house, but it didn't catch on.

But while dumping almost two million dollars into the ruinous iron-ore project—practical but

Edison's desk stands in the library of his laboratory at West Orange, New Jersey. On completing the building in 1887, Edison whimsically told a reporter he had stocked it with "everything from an elephant's hide to the eyeballs of a United States Senator!"

WILLIAM L. ALLEN, NATIONAL GEOGRAPHIC STAFF

ahead of its time—he earned new wealth from improvements on the phonograph. He worked in motion pictures, which he envisioned as accompanied by sound from a phonograph. With his younger colleague W. K. L. Dickson, he introduced the use of perforated strips of celluloid for the film, and for some years took in as much as a million dollars annually in royalties from these patents.

Edison had few intimates. Even people who had worked with him for half a century and liked him rarely felt close to him. In later years, however, he struck up a friendship with Henry Ford, 16 years his junior. They took camping trips together, with Harvey Firestone and John Burroughs, bumping over dirt roads in Ford cars and followed by a retinue of servants in Ford trucks. Burroughs reported that Edison the Ascetic slugged abed until 10 a.m. and ate pie "by the yard."

He grew increasingly imperious and crusty in old age. He wanted no part of electronics, even though a discovery he had stumbled on in the 1880's, the vacuum tube—in which he had observed but not pursued his one contribution to pure science, the "Edison effect"—served as a cornerstone of the new science.

Instead, in the 1920's, he tried to develop a domestic source of rubber, settled on goldenrod, and hybridized a plant that grew 14 feet tall. This engrossed him to the end.

It took uremia, Bright's disease, diabetes, and a gastric ulcer to bring him down. He died on October 18, 1931, only 40 hours before the fifty-second birthday of the electric light.

Someone suggested that the Nation, which had already given the old man the Medal of Honor, could best show respect for his memory by shutting off all electric power for a minute or two. It did not take much reflection to realize that the country would be crippled if Edison's power-distribution system were cut off, even for sixty seconds. Within half a century his life had altered the life of his world.

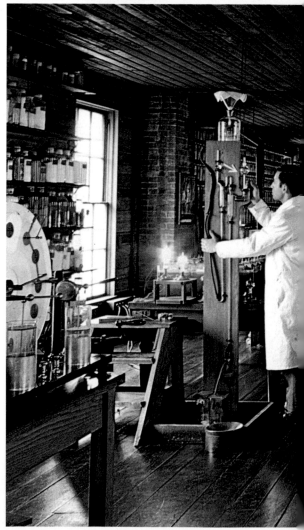

Edison's figure in bronze faces his Menlo Park laboratory, reconstructed at Greenfield Village, Dearborn, Michigan. His incandescent lamps glow in the ground floor; gas jets illumine the upstairs. Below, Associate Curator Robert Koolakian lights a replica of an early Edison bulb in the building's second floor—birthplace of the incandescent lamp and the phonograph. An array of patent models and electrical apparatus fills shelves and tables.

Herman Hollerith (1860-1929)
He made cards count—before the Computer Age

HERMAN HOLLERITH
By Tee Loftin Snell

WHEN HERMAN HOLLERITH started out for a supper date one summer Sunday in 1880, machines to read, count, and sort cards full of holes hadn't crossed his mind. He was simply a highly ambitious 20-year-old engineer a year out of Columbia College, walking across N Street in the Georgetown section of Washington, D. C., to visit a family that included a pretty girl.

The family also included her father, Dr. John Shaw Billings, a specialist at the Census Office where the gifted young engineer, seventh child of a German immigrant, had begun his first job some months before. Hollerith worked on statistics. The job had little connection with engineering, but it did pay $600 a year.

"He no doubt looked like deliberate calm itself when he knocked on the Billingses' door," says Virginia Hollerith, his youngest daughter, nearly a century later. She and her two sisters, Nan and Lucia, still live in a Georgetown mansion "designed by Father's orders to look like a factory, not pretty, in the hope of holding down tax assessments." And they tell their father's story of inventing the Hollerith Electric Tabulating Machine—work that brought "Do not fold, spindle, or mutilate" like an eleventh commandment into daily life, and led to computer systems guiding men to the moon and back.

Proper Washington society of the 1880's lived along the horse-drawn trolley line in Georgetown, the old village on the Potomac about a mile from the White House. There young Hollerith rented a room across the street from Dr. Billings. He joined the boating club, bought formal clothes, danced and dined at social events where young bachelors were in demand.

Walking past the Billingses' old row house at 3027 N Street, one can easily imagine the scene as Herman Hollerith sat with the family eating Mrs. Billings's chicken salad and drinking tea. "The two men talked shop," Virginia Hollerith reports. Dr. Billings complained about the years required to count census data. Tallying by marks

with dip-pen and ink worked fine in 1790 when there were only four personal questions for four million people. But for fifty million or more answering scores of inquiries? A faster method had to be invented or a new census would start before the old one had been counted!

Dr. Billings's conclusion might have been stated as casually as asking for lemon for his tea: "There ought to be a machine for doing the purely mechanical work. . . ."

Young Hollerith apparently paid little attention to the pretty girl or anything else that evening except, as he later wrote, talking the matter over and letting his imagination, mechanical gifts, and ambitions run with the ideas. When he walked back to his room, he might well have pondered Dr. Billings's suggestion of converting all answers to census questions into notches on cards, one card for each person. A notch's location would indicate the response.

For example, a nick on the right edge of the card would mean "male," while one on the left would mean "female." A machine could quickly "feel" each card and register on one counter those with right-side notches and on another counter the left-side notches, thus sorting and

Census clerk in Washington, D. C., slips a punched card into a Hollerith tabulating machine during the 1890 count. Each hole in the card permits a pin to slip through into a socket below, completing an electric circuit and registering one unit on the corresponding dial. At left, Herman Hollerith, Jr., resets a counting dial on one of his father's early tabulating machines at the Smithsonian Institution.

ARTHUR LIDOV; COURTESY THE HOLLERITH FAMILY (OPPOSITE); JAMES A. SUGAR, SMITHSONIAN INSTITUTION

Like patient beasts of burden, railroad cars line tracks glowing golden in the setting sun at the freight classification yards of the Santa Fe railway at Kansas City, Kansas. Here a computer sorts thousands of cars daily into trains bound for destinations across the Nation. As the inclined "hump" (bottom left) rolls each car toward the maze of tracks, the computer reads the destination

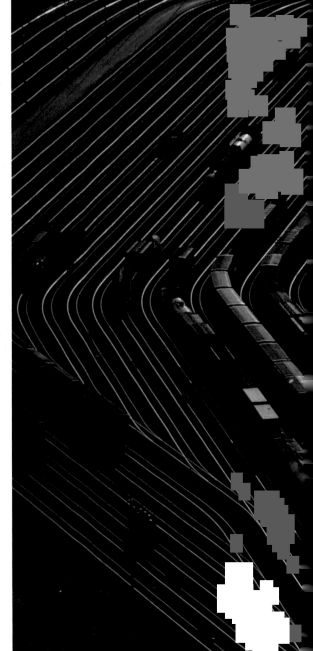

from information stored in its memory. Scales weigh the rolling car, and the computer automatically sets switches and brakes to send the car into the correct line at coupling speed. An electronic scanner (center left) identifies a rolling car from information coded on colored plastic strips on the car's side. Colored lights on a panel (upper left) in the computer room verify the working of the highly intricate system.

NATIONAL GEOGRAPHIC PHOTOGRAPHERS BRUCE DALE (OPPOSITE, CENTER) AND EMORY KRISTOF

totaling, fast and accurately, the male and female population.

If Hollerith could figure out a way to make such a machine and card system, and power it with electricity, it would surely revolutionize statistical work at the Census Office—perhaps even reward him with a fortune!

Money-consciousness, not poverty, colored Herman's upbringing. When he was 9, his widowed mother moved from Buffalo to New York City to support her children by making women's hats. Her husband, once a high-school teacher in Speyer, Germany, had made a comfortable living from renting farmland that he owned in Minnesota.

The family photograph album shows the future inventor at 9. With eyebrows and mouth tensed into dissonant slants, he exudes a strong emotion. "Maybe he's despising those baggy knee britches he's wearing," remarks his son Herman, Jr. "When my two brothers and I were growing up, he went with us to pick out our clothes and he always opposed our having knickers. He held strong opinions. He usually expressed them in a mild tone of voice, but he didn't give them up easily."

At 10, the inventor-to-be jumped from a second-story school window one day to protest spelling lessons. His mother finally had to hire a tutor for her pre-teen dropout. Just after his 16th birthday he entered Columbia College, and completed a mining-engineering course in three and a half years.

Suddenly now, in his first job, the urgent need for a machine to handle masses of census information began to absorb him. During the next three years he learned all the details of census tabulating; he scribbled machine designs on paper, made a small model.

One of the Billings daughters later wrote, "I ... remember the first little wooden model which Herman Hollerith brought over to our library many evenings while they were puzzling their brains over its adaptation. Father had no

mechanical gifts—so the entire credit is Mr. Hollerith's." Dr. Billings told him, Herman recorded later, that he had no interest in working out the machine or financing it, but he encouraged the younger man. And it seems that a romance never developed between Mr. Hollerith and the pretty girl at the chicken-salad supper.

In 1882 Hollerith accepted an invitation to teach at the Massachusetts Institute of Technology. He left his now $900-a-year job for the academic life where he would have more time for his own experiments, advanced equipment, and contact with other engineers in the new, rapidly expanding field of applied electricity.

"While at Boston," he wrote years later, "I made some of my first crude experiments. My idea at that time was to use a strip of paper and punch the record for each individual in a line across the strip. Then I ran this strip over the drum and made contacts through the holes to operate the counters. This you see gave me an ideal automatic feed. The trouble was however that if for example you wanted any statics [statistics] regarding Chinamen you would have to run miles of paper to count a few Chinamen."

Player pianos would popularize similar rolls of punched paper 20 years later; French weavers had used them for automatic-design weaving since the mid-1700's. Hollerith left no sign that he knew the details of weaver Joseph Jacquard's punched card, adapted from the paper roll, or the use that Charles Babbage, British mathematician, had made of the punched-card idea in the 1830's for an analytical "engine" or computer he never finished.

It took a trip to the American West for Hollerith to see how he should record bits of information on his cards—for notches in the edges hadn't worked well. Herman Jr., a retired mechanical engineer, tells the story.

"After a year at MIT, my father came back to Washington for a job in the Patent Office. Very handy subject to know about if you're planning to invent something. He quit after seven

Computers help men, and vice versa: Specialists at the Boeing Company watch an airplane take shape at a computer's instructions. Mathematical formulas from an electronic brain control the movements of a pen on the drafting table. Below, a computer-directed milling machine shapes aluminum for the wing of a jetliner—and bores its human companion. His job: monitoring the equipment and clearing away metal shavings after putting it into operation.

months and went West. We think he went to St. Louis to work for a railroad company while he was inventing a new railroad air brake, controlled by electricity. At that time in the West, so many train robbers posed as passengers that the Government asked the railroads to keep a record of everybody aboard. To do this, the railroads used tickets as identification cards. When the conductor first collected a ticket, he punched holes in it by those items that described the passenger in a printed list. This was called, I believe, a punch-photograph.

"I often heard my father tell about watching the conductor use a hand punch to mark his ticket—light hair, dark eyes, large nose, medium height—and realizing suddenly that his tabulating cards should be punch-photographs of each person in the census."

In a few weeks, the inventor was ready to put together a prototype of his electric data-processing machine. All he needed was money.

From Washington, where he had set himself up as "Expert and Solicitor of Patents," he sent a letter in 1884 to a brother-in-law who manufactured silk in New York. He thought he could make his machine for $1,500. He asked for some help. "I am determined to make a desparate effort," he wrote with his effective if defective spelling. The next year, he asked for more.

"No," says Virginia Hollerith, "apparently no one in the family, except maybe his mother, believed enough in Father's machine to give him any more money. He felt they let him down, and he had little to do with his brothers and sisters for the rest of his life. We think that a number of people helped him with small amounts."

Three years of struggle, and his electric tabulator was ready for a trial run. He persuaded Baltimore officials to let him tabulate death records in 1887. Armed with his train conductor's hand punch, he recorded age, sex, cause of death into grid squares printed on thousands of cards. In days he and his battery-operated machine had completed work that would have

NATIONAL GEOGRAPHIC PHOTOGRAPHER EMORY KRISTOF

occupied clerks with tally sheets for weeks.

A long, discouraging year and a half went by. At last, in January 1889, the Office of the Surgeon General of the Army hired him to tabulate some statistics. This time, instead of ordinary punches, he used another of his inventions—a keyboard punch machine.

He then set about energetically and methodically to improve his chances at winning the big prize: the contract for tabulating the 1890 U. S. census. He demonstrated his work at a convention, tabulated New York City's death statistics, and took his machine to the Paris Exposition. From the French he won his first big public acclaim, a gold medal award.

By autumn, he set up his tabulating equipment for the census contract competition. Rival methods named for their inventors, Hunt and Pidgin, had clerks with tally sheets, colored papers and inks for transcribing information. Hollerith and his machine finished about the time second-place Pidgin reached the midway point.

Today the Smithsonian Institution displays one of the machines Hollerith worked feverishly to assemble for the census. He began with an order of 30 duplicates of his handmade machine, all with interchangeable parts, from the Western Electric Company. Each cost $1,000. Millions of cards—the size of an 1890 dollar bill to fit existing file drawers—began arriving.

Only six weeks after census enumerators had sent in filled-in forms for each American family, a rough count by the Hollerith machines showed a population of about 62.5 million. For a country accustomed to learning results after a wait of years, the speed and accuracy could hardly be believed.

Hollerith went on carefully tending and improving the machines he had rented, rather than sold, to the Government. Whenever he could synthesize a useful new fact—by combining the information from two or more card holes—he rewired the registering devices to produce it.

Up to a hundred women by day, and as many men by night, punched information into cards and put millions of them one by one through the machines. Statistical totals raced out. For the first time, officials had current facts as they studied budgets for schools, pension accounts for Civil War soldiers, the occupational makeup of the national economy.

With his invention a success, 30-year-old Herman Hollerith married Lucia Talcott of Washington, pretty daughter of a civil engineer. Businessman as well as bridegroom, he took her with him to Europe in the autumn of 1890, and there he promoted his "Census machine."

During the next decade, the census facts of Canada, France, Italy, Germany, Austria, and Russia—the first census Russia had ever taken —went through the Hollerith machines.

For the farm statistics in the 1890 census, Hollerith had altered some of his machines to enable them to add bushels of corn, acres of land. By linking several dials together he made it possible to cope with addition in units as large as hundreds of thousands.

Now the Hollerith punched-card tabulator, with its built-in adding machine, could become the super-clerk that large businesses needed. For railroads it quickly traced freight cars scattered over the country, routed them efficiently, and figured rate schedules; for department stores it organized and kept up-to-date records of inventories; for manufacturers it did cost accounting rapidly and accurately.

Between the 1890 and 1900 censuses, and after the Census officials dropped Hollerith in favor of making their own machines for the 1910 census, he marketed his services to the business world. By 1915, hundreds of companies had rented his machines.

During the next 35 years, improved descendants of Hollerith's invention began managing vast quantities of statistics: accounts, payrolls, and intricate calculations for banking and insurance as well as a variety of Government programs—including atomic-bomb making. In the

First-graders in McComb, Mississippi, practice arithmetic drills, guided via telephone by a computer that serves seven elementary schools. The machine greets each child by name, and offers encouragement. The harried housewife using an adding machine to check her computer-processed bank statement may find the electronic brain less sympathetic.

NATIONAL GEOGRAPHIC PHOTOGRAPHERS EMORY KRISTOF (ABOVE) AND BRUCE DALE

late 1940's, complex successors to his machine provided a basis for the design of an electronic, or vacuum-tube, digital computer, still using punched cards for taking in information but with electronic storage units, or memory. In the 1950's, transistors took the place of vacuum tubes, bringing miniaturization. And by 1970 a new clan had been born, using not electronic impulses but laser beams to record data on plastic strips.

Hollerith himself took a much less active part in the data-processing business after 1911. That year, he sold his Tabulating Machine Company for an undisclosed sum—perhaps $2,000,000— and turned most of his attention to enjoying the good life for his last 18 years. His millions

sweetened the bitter taste of a defeat in the courts when he tried to strike at what he keenly felt was the worst kind of patent infringement —from Government machine shops.

In 1924 the name of the company carrying forward his inventions was changed to International Business Machines Corporation. For years he had ignored its high-energy, sales-minded president, Thomas J. Watson, Sr., and he ignored the whole company now by refusing to invest a penny beyond some stock he had retained when he sold his own company.

"If he had put $200,000 into the new stock of IBM," muses his son Charles, a retired engineer-inventor-executive, "the money would have increased to about a *billion* dollars today."

Twin television towers sprout a hundred stories up on Chicago's John Hancock Center. Engineer Fazlur R. Khan and his team used a small computer for preliminary design and preparing punched cards that enabled large computers to determine stresses within the structure. This proved the unusual design economical — it required about 60 percent as much steel per square foot as conventional skyscrapers.

Compared to the startling performance of electronic or solid-state computers, Hollerith's machine has the sophistication of an old cat's-whisker receiver in today's radio industry. Nevertheless, it was the first working invention of modern information technology, a new industry which by 1970 was adding twelve billion dollars a year to the United States economy.

And during the time that the value of IBM stock would have increased some 5,000 times, the uses for data-processing machines increased by astronomical numbers.

The list would fill 40 volumes or so the size of this book: teaching arithmetic or how to fly a jet; combing police records to spot a criminal; simulating skyscrapers, traffic interchanges, airplane designs, business programs, moon trips; tallying sales on the stock exchange; guarding the country against attack.

Ask friends and neighbors if they ever heard of Herman Hollerith, and it will be unusual if anyone says "yes." But in Europe his name still clings to the punched card. Technical articles describe much of the available computing equipment in Central and Eastern Europe today as "the Hollerith System." And in 1965 his daughter Virginia saw Hollerith machines handling the bets for the Irish Sweepstakes.

Herman Hollerith, with 38 patents in his name, enjoyed his success and his money. He lived with his wife and six children in handsome Georgetown houses, always kept at least one automobile — once he had four on hand — owned boats and a yacht, country acreage, the newest electrical appliances.

It was Herman Hollerith's good luck to have his special talents meet a profound idea for handling an urgent need. Fortunately he lived in an age well advanced enough in technology to make his invention possible. He pressed forward with a calculated, tenacious effort, moved by a desire he expressed simply in 1895: "... If only I have enough money, I could lead a pleasant life and be quite proud."

N.G.S. PHOTOGRAPHER EMORY KRISTOF

Boeing 727 climbs through an air traffic controller's
radar display screen in a double exposure made at
John F. Kennedy International Airport, New York.
Computer-generated symbols beside blips show
flight data. Bottom, a computer pushed colored wires
through a plotting board and clipped them to length
for a three-dimensional graph—a dramatic way to
show traffic patterns, population, and other statistics.

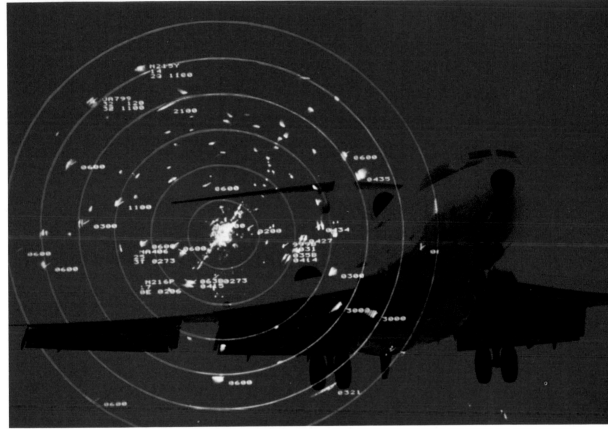

N.G.S. PHOTOGRAPHERS BRUCE DALE (ABOVE) AND EMORY KRISTOF, SPATIAL DATA SYSTEMS, INC., GOLETA, CALIFORNIA

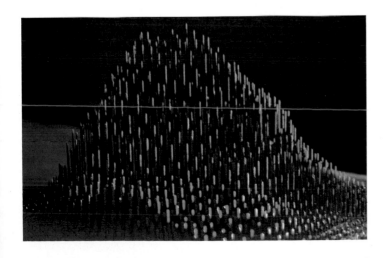

A Century Turns: America Leaves the Past

8

"D O YOU THINK . . ." a young southern matron sat rocking on her front porch one afternoon, chatting with a neighbor in the comfortable way of small-town sociability. "Do you think"—she approached the strange idea as warily as the strange word—"we'll ever ride in one of those au-*tom*-o-biles?"

Such speculations stirred the placid talk of porches and parlors, picnics and church socials of the "good years," "the good old days." Since 1860, men in France, England, and Germany had been developing the internal-combustion engine. In 1887 the teen-age sons of Mannheim engineer Karl Benz had borrowed one of his horseless carriages to take an 85-mile joyride, "thus," as an English authority observes, "presaging the anxieties of parents forever after."

As the 19th century yielded to the 20th, a different anxiety *(Continued on page 160)*

Broadway, 1900: Big-city bustle becomes a part of the American scene as a century ends. New Yorkers in this lithograph-color print take electric trolley cars—introduced in the 1890's—to skyscraping office buildings, then ride Elisha Graves Otis's elevators to top floors. The city's new melting pot absorbed immigrants by the million. Of a population of 5,000,000, four-fifths claimed either foreign birth or parentage.

LIBRARY OF CONGRESS

"Tin lizzies" of 1913, about an hour in the assembly, near completion at the Ford Motor Company outside Detroit. Bodies slide down a ramp to join chassis; an attendant drives the finished car onto a lot. While working as an engineer at the Edison Illuminating Company in Detroit, Henry Ford began building autos in a backyard shed, and in 1903 founded his company. Perfecting the assembly line, he cut production time per car from 13 hours to one. By 1924 Model T's accounted for half the world's motor vehicles. Sunday excursions became popular but tricky. At right, an E-M-F (later Studebaker) bogs down during an automobile-club tour in 1909.

HENRY FORD MUSEUM, DEARBORN, MICHIGAN (ABOVE); AMERICAN AUTOMOBILE ASSOCIATION

Inventors of the Gilded Age patent implements of questionable practicality

Of the nearly four million patents granted since 1790, a few have had enormous impact on life in America, but many inventions have left no trace. No doubt these devices seemed like good ideas at the time. R. M. Gardiner's "combined grocer's package" could serve as either a grater and slicer, or as a mouse and fly trap. Tiny spectacles invented by A. Jackson, Jr., supposedly stopped chickens

SCIENTIFIC AMERICAN, JUNE 14, 1902 (BOTTOM RIGHT); ALL OTHERS U.S. PATENT OFFICE

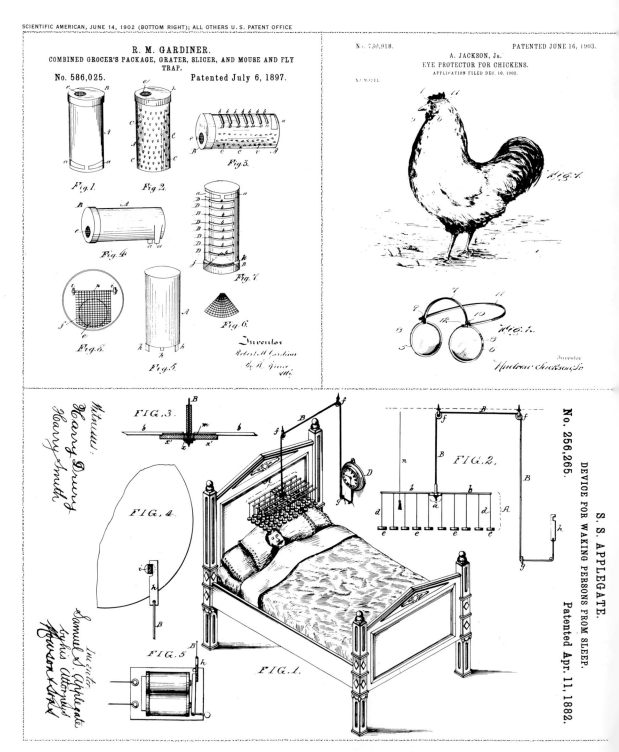

from pecking each other's eyes. S. S. Applegate described his "simple and effective device for waking persons." At the appointed hour, a frame suspended above the sleeper falls onto his face. With the "saluting device" of J. C. Boyle a gentleman with his hands full could tip his hat to a lady. Simply bowing his head activated the mechanism, thereby "automatically effecting polite salutations." The owner of T. W. Helm's "combined clothes brush, flask, and drinking cup" might have taken a surreptitious nip in his dressing room. J. B. Campbell's contraption promised to extract either vegetable or animal poisons from the body by depositing them on a copper plate at the patient's feet. A hairbrush, intended for use in public washrooms, kept its bristles recessed until fed a coin.

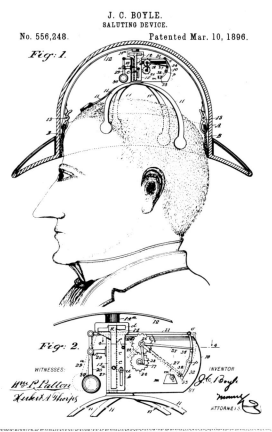

J. C. BOYLE.
SALUTING DEVICE.

No. 556,248.   Patented Mar. 10, 1896.

Fig: 1.

Fig: 2.

WITNESSES: INVENTOR

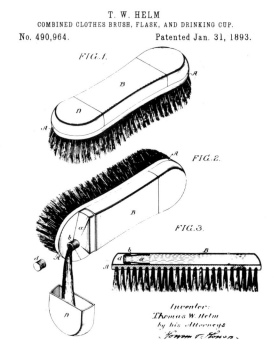

T. W. HELM
COMBINED CLOTHES BRUSH, FLASK, AND DRINKING CUP.

No. 490,964.   Patented Jan. 31, 1893.

FIG.1.

FIG.2.

FIG.3.

Inventor:
Thomas W. Helm
by his Attorneys

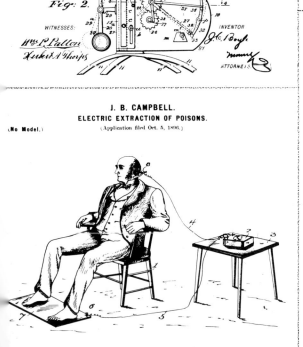

J. B. CAMPBELL.
ELECTRIC EXTRACTION OF POISONS.

(No Model.)   (Application filed Oct. 5, 1896.)

COIN-CONTROLLED HAIR BRUSH.

EASTMAN KODAK COMPANY (OPPOSITE); POLACOLOR BY INGE REETHOF

Cameras born 80 years apart capture images of their inventors: George Eastman, aboard the S.S. Gallia in 1890 (opposite), aims one Kodak camera while another records his picture. First to use roll film, the camera introduced millions to photography. Edwin H. Land (below) sits in his office with a Polaroid Land camera; his self-processing film recorded this likeness. The first Polaroid, marketed in 1948, produced a black-and-white print in 60 seconds.

beset American leaders: a fear of revolution. "Nothing," asserted Dr. Woodrow Wilson of Princeton University, "has spread Socialistic feeling in this country more than the use of the automobile"—a symbol, he meant, of the chasm between the rich who could afford such ostentatious luxuries and the poor, the out-of-work, who wondered if they could still afford to hope. And Wilson spoke just four years after Theodore Roosevelt, in 1903, had encouraged the use of automobiles and ridden in one himself.

In the West, the old frontiers had disappeared, for the pioneers had settled down to develop the country. Immigrants from Europe—more than 13 million between 1865 and 1900—too often worked for too many hours at miserably low wages. The country Franklin described so glowingly in the 18th century had changed.

A coal miner might earn $10.75 in a six-day week, a shopgirl $6, a black workman in Baltimore half as much, a field hand in Virginia $7 or $8 a month and board.

Labor-saving inventions might ruin men. In the shoe trade a skilled handlaster could turn out 50 pairs in a ten-hour day. A laster patented in 1883 by Jan Ernst Matzeliger, a young part-Negro immigrant from Dutch Guiana, could do 150 to 400 pairs; and to old-style craftsmen Matzeliger's machine said, "I've got your job."

On December 31, 1899, a New York editorial declared: "This has been a century of mechanical invention rather than of social reconstruction...." The new century would have to invent new ways of distributing its wealth.

Social inequalities, labor violence, lynch mobs, anarchist bombings: all these marked the good years, and some of the sharpest comments around came from "Mr. Dooley," an Irish barkeep in Chicago, created by journalist Finley Peter Dunne in the 1890's. When Martin Dooley saw "little boys an' girls with their dinner-pails on their arms" going to work instead of to school, "I'm th' hottest Socialist ye iver see. I'd be annything to stop it."

But he wouldn't do just anything. He opposed the use of dynamite: "Even an Englishman was niver improved be bein' blown up." He would not call new immigrants "th' scum iv th' earth," and he mocked his friends who thought America was ruined: "I'm not afraid iv me father and I'm not afraid iv mesilf." He was not afraid of any fellow citizen "too sthrong and indepindant to be kicked around."

An old confidence spoke in new accents as the United States reviewed one century of invention—and began another.

Lidov

The Yielding Sky: Exploration Leaves the Earth

9

DOWN THE AGES, the lore of invention carries a sparkling vein of pure absurdity, more or less self-evident—an endearing and humbling reminder that man's ingenuity soars with ease beyond the eccentric into the wacky, the barmy, the unworkable. And nowhere more vividly than in inventors' attempts to fly.

They hitched flappable wings to themselves, or kites to carriages. They sketched out plans for hitching geese to a trapeze. They saw glorious possibilities in steam for aerial locomotion —with wings attached to a piston.

Their early machines trembled and rocked and jerked and skittered forward, and bounced and now and then blew themselves to smithereens.

They looked up wistfully from their earthbound contraptions to the changing skies where balloonists drifted or cruised about in their gas-filled "aerostats."

Giant step toward the sky: The Wright brothers test their successful No. 3 glider at Kill Devil Hills near Kitty Hawk, North Carolina, in October 1902. During these trials they added a movable rudder controlled by the same wires that twisted, or warped, the wings of their craft. This enabled them to maintain stability and glide as far as 622½ feet, opening the way for their first powered flight 14 months later.

ARTHUR LIDOV

Toy helicopter powered by rubber bands—a gift from their father—lands in a tree above Wilbur Wright, 11, and Orville, 7, at their Cedar Rapids, Iowa, home in 1878. Invented in France several years earlier, the novelty led the brothers to build and test larger models.

THE WRIGHT BROTHERS
By Joseph J. Binns

They persevered. "Mere shape," one hopeful reminded his readers, "determines whether iron shall float or sink." Another tried—with zeal—modifying the shape of a bat.

By 1900 a Russian, a Frenchman, and a Britisher could claim they had gotten their machines off the ground under power, however briefly.

And by this time the "aeromobile" seemed a logical successor to the automobile, with its compact internal-combustion engine. The happy aeronaut would drive triumphantly into the sky.

Considering in 1898 that a flying machine might serve in the war against Spain, President William McKinley and the War Department made $50,000 available to America's leading scientist of flight, Samuel P. Langley, then Secretary of the Smithsonian Institution.

With patient diligence, recording each of thousands of modifications, Professor Langley had successfully flown his steam-powered "aerodrome" models. On October 7, 1903, he launched his man-carrying *Aerodrome* with its 52.4-horsepower gasoline motor from a catapult over the Potomac River. An eyewitness reported that it "slid into the water like a handful of mortar." The pilot was saved. Congressmen threatened an investigation into such waste of public money. Langley was heartsick.

According to the newspapers, as a young Midwesterner had noted in February 1902, innumerable less-qualified inventors had been busily building flying machines "in cellars, garrets, stables and other secret places" to win a prize of $100,000 offered by promoters in St. Louis. All of them had the problem "completely solved" except for "some insignificant detail . . . such as whether to use steam, electricity, or a water motor. . . . Mule power might give greater *ascensional force if properly applied,* but I fear would be too dangerous unless the mule wore pneumatic shoes."

A sardonic commentary, but in character. Its author, Wilbur Wright, like his brother Orville, had his own view of these problems.

"For some years I have been afflicted with the belief that flight is possible to man. My disease has increased in severity and I feel that it will soon cost me an increased amount of money if not my life. I have been trying to arrange my affairs in such a way that I can devote my entire time for a few months to experiment in this field."

So, on May 13, 1900, businessman Wilbur Wright, co-owner with his brother Orville of the Wright Cycle Company, makers and purveyors of bicycles to the American public, wrote to Octave Chanute, consulting engineer and chronicler of man's desire to fly.

Captivated by the style and sophistication of the letter and the author's sensible plans, Chanute replied immediately with reports of aeronautical efforts in France, Britain, and Australia, beginning the most important correspondence in the history of aviation.

Both Wilbur—who was born April 16, 1867, near Millville, Indiana—and Orville—born August 19, 1871, in Dayton, Ohio—had this affliction and both got it the same day.

Their father, the Reverend Milton Wright, a bishop of the United Brethren in Christ, was the carrier. In the summer of 1878 he gave the boys a flying machine, a toy helicopter invented by Alphonse Pénaud, a young Frenchman. Twisted rubber bands propelled the toy, made of bamboo, cork, and paper, through the air.

After wearing it out, the boys decided to make larger models to fly higher and farther. They discovered with chagrin that the larger the model, the less it would fly. Puzzled, they turned to simpler machines.

As they grew up, the boys tinkered with any mechanical device that came their way, and both were inveterate readers, especially fond of the *Encyclopædia Britannica.* Perhaps inspired by older brother Lorin's improvements on a haybaling machine, Wilbur designed and built a practical newspaper-folding device.

Mrs. Wright, never *(Continued on page 170)*

ARTHUR LIDOV

VICTOR R. BOSWELL, JR., N.G.S. STAFF (TOP); LIBRARY OF CONGRESS (BOTTOM); NATIONAL AIR AND SPACE MUSEUM, SMITHSONIAN INSTITUTION

Moved from Dayton and reconstructed, the Wrights' bicycle shop stands in Greenfield Village at Dearborn, Michigan. The brothers mounted airfoils on the bicycle device to compare their efficiency, and tested wing configurations in a small wind tunnel like the one against the wall. Reproductions of experimental wing ribs lie on the bench at right.

Glider launched in 1911 stands on tail and wingtip (opposite), flipped by a gust at takeoff. Next day, October 24, Orville piloted the craft to a 9.75-minute soaring record that lasted a decade. Ten years earlier, another glider (left) with a 22-foot wingspan rises at the end of its tether. The brothers chose the Kitty Hawk area as a test site because of its sandhills, winds, and lack of trees or other obstructions.

LIBRARY OF CONGRESS

December 17, 1903, near Kill Devil Hills: As Wilbur watches tensely, Orville Wright pilots their Flyer for 12

history-making seconds—the first sustained, controlled, powered flight of a heavier-than-air machine.

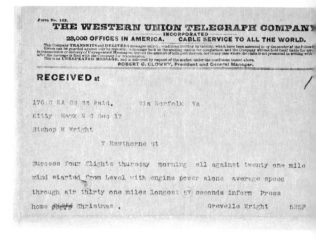

robust, died in 1889; and after the two older brothers married and moved away, a tight family bond developed among Bishop Wright, Wilbur, Orville, and younger sister Katherine. These two brothers never married.

Although everything about the home encouraged their intellectual interests, neither Wilbur nor Orville got around to taking a high-school diploma. In 1892 they opened a bicycle business in Dayton. The tall, plainly dressed Wilbur and his shorter, mustachioed, dapper brother quickly built up a reputation for integrity. In 1895 they began making their own bicycles: first a custom model, then a less expensive one, then the mass-market Wright Special, which sold for about twenty dollars.

A year later they learned of Otto Lilienthal's glider experiments in Germany and began reading everything they could find about flying. By 1899, having exhausted the resources of Dayton, they wrote to the Smithsonian Institution for advice. The reply suggested Samuel P. Langley's *Experiments in Aerodynamics,* Chanute's *Progress in Flying Machines,* and other works. These made them aware of the current state of the flying art, and a decade of correspondence and friendship with Chanute followed.

Reading gave the brothers a good deal of misinformation, but they worked with a crucial observation as well: that large soaring birds kept their lateral balance by twisting the tips of their wings. They thought of applying this method to gliders, and Wilbur hit upon a technique of twisting or "warping" the wings while playing with a pasteboard inner-tube box.

To test this, the brothers built a biplane kite of 5-foot wingspan in 1899. It had a tail plane that moved up or down with the shifting of the wings—a primitive elevator to control a climb or a dive. When this model proved the warping method workable, they decided to go on, preferably in an area with consistent winds and sandhills free from trees.

By September 1900 they had weighed Weather

Orville circles the Wright Military Flyer above the parade ground at Fort Myer, Virginia, during trials for Army officials in 1909. Above, the telegram notifying their father of the Wright brothers' triumph of 1903 acquired errors in transmission: 57 seconds should read 59 seconds, and Orevelle should read Orville.

LIBRARY OF CONGRESS (TOP); NATIONAL AIR AND SPACE MUSEUM, SMITHSONIAN INSTITUTION

Bureau reports and chosen Kitty Hawk, on the Outer Banks of North Carolina, as a testing ground for their first full-sized glider—a biplane with a 17-foot wingspan and a horizontal rudder in front. Because the white sateen wings measured only 165 square feet in area, the brothers usually flew it as a kite, but they risked a few glides two or three feet above the sands.

They returned in July 1901 to the steeper slopes at Kill Devil Hills with a new and larger glider. Its wingspan of 22 feet and area of 290 square feet promised more lift. The wings in cross section had a marked curve from leading to trailing edge—a camber of 1 in 12—as the best authorities recommended. The pilot would lie prone in a "hip cradle"; when a wing dipped and he automatically shifted his weight (as a bicycle rider does) to balance the craft, wires attached to the cradle would warp the wings.

Tested in flight, the glider was not very successful. Reducing the camber of the wings improved it somewhat, but the brothers saw that the calculations they had relied on, Lilienthal's and others', were in error. As Wilbur put it: "Having set out with absolute faith in the existing scientific data, we were driven to doubt one thing after another, until finally, after years of experiment, we cast it all aside, and decided to rely entirely upon our own investigations."

Settling down to lab work in the bicycle shop, they built a small wind tunnel and tested more than 200 types of wing surfaces. The results, far different from those in print, were so accurate that modern figures, obtained by the most sophisticated instrumentation, offer only slight refinements. Without fuss, the brothers were making scientists of themselves; their research during 1901 and 1902 made it possible for them to design an airplane that would fly.

They incorporated their results in a glider designated simply No. 3, tested in the fall of 1902. No. 3 had a wingspan of 32 feet, a wing area of 305 square feet, and a very shallow camber to the wings. It also had a fixed double fin

in the rear. This machine performed well except for sporadic trouble in deep-banked turns.

Orville solved this in a sleepless night after drinking too much coffee, and a single movable vertical rudder—linked with the wing-warping mechanism—replaced the fixed double fin. No. 3 now banked smoothly and the Wright brothers had the first practical glider in history, a machine in which they made almost one thousand *controlled* glides, several of more than 600 feet and many into winds stronger than 35 miles per hour.

Deeply impressed, a spectator from Kitty Hawk had remarked, "All she needs is a coat of feathers to make her light and she will stay in the air indefinitely."

With the control of flight assured by their co-ordination of wing-warping (for lateral stability), elevator (for climb or dive), and rudder (for turning), the brothers applied for a patent on this system in March 1903. Between examiners' questions and office routine, the paper work progressed more slowly than the flying art: Patent No. 821,393, for "certain new and useful Improvements in Flying-machines," was granted May 22, 1906.

By the time Wilbur and Orville filed the application, they had made themselves expert pilots. Confident in their skill, they tackled the problem of powered flight. They had expected to use an automobile engine; finding nothing light enough for the power they needed, they produced a small 12-horsepower motor.

Turning to marine engineering for data on the screw propeller, they found its exact action "after a century of use, was still very obscure." They settled down in the shop to experiment and argue. Through all their work with flying machines, they debated and discussed and challenged each other freely. Thrashing out the mysteries of the airscrew was a bit unusual: "After long arguments we often found ourselves in the ludicrous position of each having been converted to the other's side. . . ."

The first Wright Flyer, named for one of their

Rescuers hurry to the aid of victims in a crash at Fort Myer on September 17, 1908. Lt. Thomas E. Selfridge, a passenger, died in the accident, becoming the first airplane fatality; Orville Wright, seriously injured, recovered. Below, Wilbur performs for King Alfonso of Spain at Pau, France, in February 1909. The King wanted to go up, too, but his Queen and his cabinet made him promise that he would not.

bicycles, was ready by autumn, 1903. A biplane, it had skids like sled-runners for undercarriage, flattened wings (a camber of 1 in 20), a wingspan of 40 feet 4 inches, a wing area of 510 square feet. It had a biplane elevator in front and a double rudder in the rear. The engine, with a chain transmission like a bicycle's, drove two propellers in opposite directions.

Tossing a coin "for first whack," Wilbur won the privilege of trying the first flight on December 14 at Kill Devil Hills. The Flyer plunged into the sands three and a half seconds after takeoff because of an overcorrection of the elevator, and sustained minor damage that was repaired over the next two days.

On the morning of the 17th, "conditions were very unfavorable as we had a cold gusty north wind blowing almost a gale." Under ordinary circumstances they would have waited for safer weather, but they were determined to get home for Christmas dinner with the family and they had faith in their calculations and their skill.

It was Orville's turn. After warming up the engine, he started forward into a 27-mile-an-hour wind. The machine lifted off its wooden-track "runway" after 40 feet and flew erratically for about 120 feet before darting to the sand. Orville summed up this 12-second flight a decade later: "the first in the history of the world in which a machine carrying a man had lifted itself by its own power into the air in full flight, had sailed forward without reduction of speed, and had finally landed at a point as high as that from which it started."

The brothers made three other wind-buffeted flights that morning, the last, by Wilbur, covering 852 feet in 59 seconds.

It was noon on December 17, 1903, and two young men with determination, a passion for experiment and scientific method, and an expenditure of less than $1,000, had realized man's most persistent and primordial dream.

But after decades of reported "flights" and failures by scientists, half-baked inventors,

sensation-seekers, and assorted cranks, the press was wary. The next morning most of the newspapers in the United States carried no account; according to current research, only 17 saw fit to publish the story.

Over the next two years the brothers worked to develop the airplane as a practical method of transportation. A friend, Torrence Huffman, let them use a 70-acre pasture between Dayton and Springfield. They made more than 120 flights, mastering banks, turns, circles, figure eights. The press ignored them.

In 1905 the Wrights, believing their airplane to have military value, wrote to the Army describing their flights and offering to produce aircraft with agreed-upon specifications at a price to be negotiated. Forwarded through channels, their letter reached the Board of Ordnance and Fortifications. This worthy group, badgered for money by too many mad geniuses, sent back their form reply: "... the device must have been brought to the stage of practical operation without expense to the United States...."

After fruitless negotiations with the British Government, who knew the Wrights had a practical airplane, and urged on by Chanute, the brothers renewed their offer to the U. S. Army. They promised a machine to carry a pilot and fuel for 100 miles, a minimum flight of 25 miles at 30 miles an hour, and trial trips for proof.

The Army repeated verbatim the opening paragraphs of their form letter and demanded drawings to prove that the 1905 Flyer could fly.

The Wrights gave up, Wilbur stating that it was bad business practice to try "to force goods upon people who did not want them."

For another two and half years they continued to improve their invention.

They kept their sense of humor. In fact, they planned a practical joke on the military that would equal the dream revenge of any citizen irritated by bureaucrats. It was 1907, the year of the Jamestown Exposition, and the Navy was arranging a grand fleet review at Hampton Roads

NATIONAL AIR AND SPACE MUSEUM, SMITHSONIAN INSTITUTION

Nearly six miles high, U. S. Navy F-4 Phantom jets sweep over North Carolina's Outer Banks where, 65 years earlier, the Wright brothers first proved powered flight possible. Farther frontiers of space beckoned later pioneers like Robert H. Goddard (left). He stands beside his liquid-fuel rocket—the world's first—before its flight in 1926. Another of his many inventions: the rocket-firing bazooka.

Robert H. Goddard (1882-1945)
He steadied our first steps toward the moon

in April, before the high dignitaries of the Nation. The Wrights decided to equip an airplane with pontoons: to appear suddenly, circle the fleet, and fly away—leaving officials to explain to President Theodore Roosevelt how the Navy had been buzzed by a machine not yet brought "to the stage of practical operation."

Fortunately for the Navy and the Army, a mishap damaged the Flyer while the Wrights were preparing for the stunt.

Finally, in early 1908, the brothers signed a contract with the Army, as well as one with a French company—both on terms dictated by the Wrights. Wilbur went to France in May to dem-

onstrate their new "A" aircraft and Orville stayed to conduct the Army tests at Fort Myer, Virginia.

European aviation, checked by the death of Otto Lilienthal in 1896, had begun to stir again after news of the Wrights' glider experiments reached France. Octave Chanute lectured in Paris in 1903 on his friends' achievements, and the French took the lead in European efforts to conquer the air. Alberto Santos-Dumont of Brazil won world renown on November 12, 1906, at Paris, when his "14-*bis*" flew 700 feet—and Wilbur had covered 24 miles a year earlier!

A skeptical audience of French air enthusiasts, inventors, and reporters gathered at the racetrack of Hunaudières on August 8, 1908, for a first test of the Wright A biplane. Wilbur took off, made two easy circles over the course, and landed perfectly. He had flown as gracefully as a bird flies. The audience was stunned, then wildly elated. Wilbur Wright—intelligent, modest, unassuming—was a French national hero.

On September 3, Orville began the Army tests. After ten successful flights he and his passenger, Lt. Thomas E. Selfridge, crashed on September 17. Selfridge died of a fractured skull, but Orville escaped with a broken leg and four cracked ribs. The publicity finally convinced the American public that the airplane was real.

When Orville recovered, he and Katherine joined Wilbur in Europe for a triumphant grand tour. They remained unaffected through it all.

The Wright Company, organized with Wilbur as president when they returned home, prospered and grew tremendously. Then Wilbur died suddenly of typhoid fever on May 30, 1912. Of his loss, Orville said little. He sold the company in 1915, uninterested in running a great business enterprise.

Orville survived until January 30, 1948, the revered elder statesman of the air. The era of learning to fly had long since ended; the age of flight to the moon was beginning. As Bishop Wright had said of his sons on December 22, 1903: "About equal credit is due each."

NATIONAL GEOGRAPHIC PHOTOGRAPHER EMORY KRISTOF (ABOVE); COURTESY ESTHER C. GODDARD

**Breaking the bonds of gravity:
an ages-old dream of flight comes true**

Evolution of flight: from batlike glider to space shuttle. German genius Otto Lilienthal made some 2,000 flights in the 1890's, sometimes soaring as far as a thousand feet. His control methods paid tribute more to his incredible courage than to his inventiveness. With straps supporting his arms,

he grasped a crossbar and swung his dangling hips and legs to counter gusts of wind that threatened to dash the fragile craft to earth. Finally perfecting in a rush of wings the art and science of flight, man then mastered rocket propulsion and reached past the limits of the earth's atmosphere. Only 78 years elapsed between Lilienthal's first glider flight in 1891 and the landing of American astronauts on the moon. Now undergoing wind-tunnel tests, the two-stage space shuttle—designed to lift off vertically toward orbit, and to land like an aircraft on its return to earth —promises to provide a stepping-stone to the planets.

ARTHUR LIDOV

The Eloquent Air: Triumphs of Electronics

10

"TO ALL CREATION . . . Hear, O ye Heavens, . . . and harken ye people from afar. . . ." For the first time, on February 12, 1931, a Pope spoke directly to the world by radio—*"ipsa mira sane ope Marconiana,* this truly wonderful Marconian invention." Monsignor Francis J. Spellman translated for Americans the ringing Latin as Pius XI addressed mankind.

In that decade families all over the earth were accepting as daily routine the living voices and music from the air: a thing men had imagined since antiquity as revealing only the supernatural.

And already pioneers of television were developing the instruments that would turn the stuff of equally ancient dreams and visions— moving images—into matter-of-fact reality.

Edison's light bulbs played a crucial role in the technology of these astonishing devices, but Europeans contributed the science that made

Born to tinker, young Lee de Forest builds a locomotive of barrels and packing crates at his Alabama home. As a man he patented hundreds of electronic devices, from radiophones for the U. S. North Atlantic Fleet in 1908 to instruments that measured a World War II airplane's speed, course, and altitude. Long before he married silent-screen star Marie Mosquini, he devised equipment that made "talkies" possible.

STANLEY MELTZOFF

them possible. Michael Faraday, "prince of experimenters," had pictured "lines of force" extending into the air from his magnets, and later from electrically charged bodies also.

The great and witty James Clerk Maxwell gave these notions mathematical form in the equations which still bear his name—and which predict the behavior of magnetism and electricity. He deduced the nature of light as waves of electromagnetic radiation, and the existence of invisible waves as well.

In two years of arduous experiments, 1887-1888, a German physicist named Heinrich Hertz managed to produce such waves at a lower frequency than light's, detecting their presence by watching an electrical spark brighten or dim.

As Franklin had linked a parlor curiosity with storm bolts, his successors had extended it to the daily blaze of sunlight, the chilly glitter of stars, and unseen kindred waves everywhere.

Within ten years young Guglielmo Marconi had adapted existing transmitters and receivers, invented a grounded antenna, and sent telegraphic messages 25 miles without wires, by the wave frequencies now called radio.

Wherever Western technology flourished, inventors took up wireless with curiosity and delight, with dreams of new patents and great fortunes, of fame and glory. Soon there were almost as many new transmitters and detectors (receivers) as there were ambitious men—with claims and quarrels for nearly all of them.

Who was first with what? One scholar concludes that "it is senseless—if not impossible —to establish priorities" in much of this work. Yet some names stand out, like Lee de Forest's or Vladimir Kosma Zworykin's.

Consequences stood out early and clearly. For a decisive new factor in World War I, the commands of wireless could reach the captains of battle fleets at sea. The bitter messages of combat clamored for better instruments, and the basic technology of radio took shape in reply.

With the return of peace, amateurs across the United States—the ham operators—chatted cheerfully with each other; and in 1920 a ham in Pittsburgh took to playing phonograph records for neighbors who owned crystal sets. Commercial broadcasts followed, from stations KDKA in Pittsburgh and WWJ in Detroit. From a permanent hookup established in 1923 came the National Broadcasting Company of 1926, and the spread of network radio.

A generation of boys learned electronics on more-or-less homemade wireless sets, tinkering with the reliable telegrapher's key or temperamental cat's-whisker receivers in innumerable American backyards. Many would grow up to be men of science and invention.

Day by day, sick, lonely, aged people found diversion at the turn of a dial. Music lovers all over the United States heard arias from the stage of the Metropolitan Opera House, or Grand Ole Opry from Nashville, Tennessee. Citizens of the 1930's found reassurance in the fireside chats of Franklin D. Roosevelt and menace in shouts of *"Sieg Heil!"* Housewives escaped into soap opera while children waited for crime fighters to capture bad guys.

Today, television brings the world to the family room, introducing the grin of the entertainer or the grimace of the wounded.

And every hour on the air mocks an old toast given at the Cavendish Laboratory, in England. There, in the 1890's, physicist J. J. Thomson discovered the electron, first known subatomic particle. Without the flourishing electric-light industry, as he was fond of pointing out, he could not have managed his investigation—for his work demanded good vacuum pumps and, to meet the needs of the light-bulb makers, manufacturers were ready to supply them.

Thanks to the commercial success of Edison's inventions, electronic science could succeed in its turn. But for some years the electron remained a laboratory curiosity. At annual dinners the Cavendish men would chorus: "The electron: may it never be of any use to anybody."

Early television: Nipkow disk receiver, patented in 1884, re-creates a transmitted image mechanically. The viewer watches a flickering bulb through a magnifying lens and a pierced, spinning disk. Square holes near the edge of the disk divide the light into bright-and-dark lines; the viewer's eye fuses the lines into a continuous image. The device proved impractical, but it established a principle underlying all TV receivers: assembling a sequence of bright-and-dark lines rapidly enough to create an image.

Television pioneer Philo T. Farnsworth examines an "image storage and display tube" in his laboratory at Fort Wayne, Indiana, in 1957. Much costlier than standard picture tubes, his invention can display radar or TV images in bright sunlight. Using such tubes, controllers in airport towers can look directly from the daylight outside to their radar screens without waiting for their eyes to adjust.

VICTOR R. BOSWELL, JR., N.G.S. STAFF (UPPER); BOTH SMITHSONIAN INSTITUTION

Lee de Forest (1873-1961)
Restless explorer of electronics

LEE DE FOREST
By Howard J. Lewis

ONCE THE TELEPHONE had been invented, the glittering prospect of wireless communication appeared simultaneously to a small army of American and European inventors, and their race to develop the first workable system took on the scrambling urgency of a gold rush. In this race a band of hurrying adventurers outdistanced scientific understanding, hoping to stake out a valuable patent claim.

At the start, the odds favored the Europeans with their more extensive grounding in the underlying fundamental research. But by hindsight one can see that none was better suited to plunging ahead in the race than Lee de Forest. His rare combination of darting intelligence, steadfast endurance, and burning ambition set him among the front-runners.

Hindsight in the case of Lee de Forest has help from a proud and voluble autobiography written at the age of 77. It reveals—not always by intention—the history of a gifted youth challenged in mind and body by a loving but authoritarian father and isolated by circumstance. At the age of 13, he already felt "that I was, not that I was to be an inventor!" Perhaps in invention he gained self-assurance from public acclaim.

He was born on August 26, 1873, in the parsonage of the Congregational Church in Council Bluffs, Iowa, second child of the Reverend Henry Swift De Forest. (Lee's father and grandfather chose to capitalize the *D;* later Lee and his brother returned to the style of prior generations.) At 6 Lee, with his older sister and younger brother, was taken to the small Alabama village of Talladega. To a family descended on the father's side from Huguenot settlers of New England and on the mother's from John Alden, Talladega must have seemed a colonial outpost.

But it was there that Dr. De Forest accepted a call to become the first president of a normal school, later a full-fledged college, for the children of former slaves.

And it was there that young Lee, bewildered by his father's black wards, ostracized by the local white gentry and tormented by their children, was led by his vivid imagination to forage among the mechanical arts—poring over the stylized mechanical intricacies of the *Patent Office Gazette,* designing steam hammers and perpetual-motion machines by the hour.

As an inventor should, he also chose to express himself in construction. At an age when most boys are content to construct a working raft, de Forest later recalled with relish, he set out to build a working blast furnace.

Putting things together, he found, was closely coupled to understanding how things worked. He was 10 or 11 when he decided to find out how a locomotive reversed its direction. In the railroad yard of the Shelby Iron Works, he located a resting locomotive, and traced the linkages from the great reversing lever in the cab through all the eccentric rods, pistons, and valve rods, until suddenly the entire system was clear to him. Hugging himself with glee, he skipped home singing aloud, "Oh, I am happy; I am so happy."

Assembling a number of large packing cases, sugar barrels, paint kegs, strips of wood, a tin can, a dinner bell, Lee and his brother Charlie lashed and nailed together over a period of weeks the largest wooden locomotive ever seen in Talladega—complete with an operative reversing lever. Not only did the locomotive

Modern broadcasting begins: De Forest airs music programs in 1907 from his makeshift Manhattan studio. A telephone mouthpiece served as the microphone; a flagpole supported the antenna. Listeners in the vicinity telephoned to report reception. From about 1913 until the transistor replaced it, radio relied on his grid vacuum tube (right).

SMITHSONIAN INSTITUTION (OPPOSITE)

Patented as a "Device for Amplifying Feeble Electrical Currents," de Forest's vacuum tube contains three elements. The electrically heated cathode "boils off" negative electrons that seek the positive anode. The negative grid, controlled by the fluctuating charge of the antenna signal, or input wave, governs the electron flow like a valve. Thus electrons reach the anode in a controlled current as an amplified copy of radio signals picked up by the antenna.

impress his father, he observed, but it also attracted the respectful attention of his hitherto contemptuous white neighbors.

Perhaps even more telling was another experience, apparently of that same period. One afternoon, his father quietly took him to join a group of adults gathered around a small boxlike object. One by one the men of the village placed their ears to the hand-cranked contraption. As each listened, his face would light up with great astonishment. Then it was Lee's turn, and his eyes opened widest of all, as he heard the sound of a man's voice come out of the box. What he was listening to was one of Edison's first tinfoil phonographs. What he heard was the urging of his life's ambition.

At 15 Lee sat down at a typewriter and wrote a letter to his father:

"Dear Sir... I intend to be a machinist and inventor, because I have great talents in that direction.... If this be so, why not allow me to so study [at Yale's Sheffield Scientific School] as to best prepare myself for that profession?... I think you will agree with me about this on reflection, and earnestly hope you will act accordingly and educate me for my profession."

With genuine reluctance and imperfect grace, Dr. De Forest let his son abandon prospects of the ministry: "Well, Lee, if you positively know you want that sort of half-baked education, you may have it.... I can only say I hope you never regret the choice you are now making."

If Lee ever felt regrets, he never acknowledged them, although his years at Yale—supported by an endowed scholarship—were not marked by complete success. His diary records the keen disappointments of a student who aspired to win $50,000 by designing a novel underground trolley system, and to become editor of the campus magazine. He succeeded only in being voted by his fellow seniors as the nerviest and homeliest man in the class. Passed over in elections to the honorary scientific society of the Sigma Xi, he wrote in his diary: "... I shall show

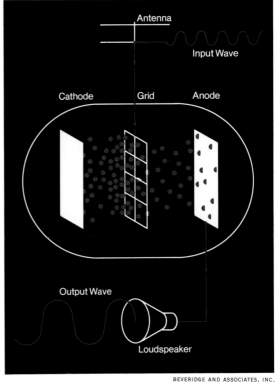

Antenna

Input Wave

Cathode   Grid   Anode

Output Wave

Loudspeaker

BEVERIDGE AND ASSOCIATES, INC.

them some day what a mistake they made. I will honor them then, and not they me!"

Nevertheless, de Forest chose to remain at Yale for a Ph.D. on the mysterious electromagnetic waves discovered in 1887 by Heinrich Hertz. During seven long months in a cold cellar laboratory de Forest fretted over the erratic nature of a new device designed to detect these waves. It was the Branley coherer, named after its French inventor, and a critical link in the wireless telegraphy system of Guglielmo Marconi. De Forest thought he could do better.

He wasted little time in setting out to prove it. Within three years of his Ph.D., the American De Forest Wireless Telegraphy Company had shown itself capable of sending the dot-dash

The way to America's heart: through her ears. Nearly 92 percent of U. S. homes had radios when World War II stopped the production of new sets. Four of every five Americans heard F.D.R.'s Pearl Harbor speech. Commercial broadcasting electrified the Nation's politics and pastimes as air became the prime medium for news and entertainment. It even gave rise to a new dramatic form—at one time 65 soap operas saturated daytime radio.

wireless messages at least six miles. The infant company procured several lucrative military contracts, but de Forest's ambition had already raced far ahead. In February 1902 he wrote: "I shall move all heaven and earth to put in at once a broad fundamental patent on *telephony* without wires by Hertzian waves."

Meanwhile, new commercial areas waited to be pioneered. And wireless pioneering in the new century required the characteristics needed to open the new continent—courage, intelligence, and stamina. Of his attempt to establish commercial communication over 180 miles of ice and frozen ground between Cleveland and Buffalo in the bitter winter of 1904, de Forest wrote: "I shall never forget the icy dreariness of that lonely location . . . the agony of raising again and again that fan aerial after sleet had piled it and the stiffened hemp halyards an inch thick in ice. . . ." He even recorded that the flapjacks were so leathery he was able to paste stamps on one and mail it to New York City.

Much of the transmitting equipment was treacherous. The only method of generating Hertzian waves was still that used by Hertz—an electrical spark. To cross great distances required great sparks and that meant powerful transformers and condensers, to store current. The condensers were liquid-tight boxes 7 feet long and 2½ feet in cross-section, built of 2-inch planks. These were filled with kerosene and lined with racks separating large glass plates, each plate with heavy tinfoil fixed to both sides. A typical series of condensers consisted of four to six of these tanks, each weighing about a ton.

The entire apparatus was likely to explode in an instant of miscalculation, smashing the plate glass and spraying kerosene over the room and its occupants. When the condensers were operating, an elbow or head that came too close often met a stunning bolt of electricity. And the sparks generated the acrid smell of ozone to mix with the always-present kerosene fumes.

Although de Forest recalled this as a period of

CBS RADIO

Sound-effects men give new depth to radio drama

RCA's first radio studio, WJY, New Jersey, 1921

SMITHSONIAN INSTITUTION

184

NATIONAL BROADCASTING COMPANY

Orson Welles meets Charlie McCarthy and ventriloquist Edgar Bergen

NATIONAL BROADCASTING COMPANY

UNITED PRESS INTERNATIONAL

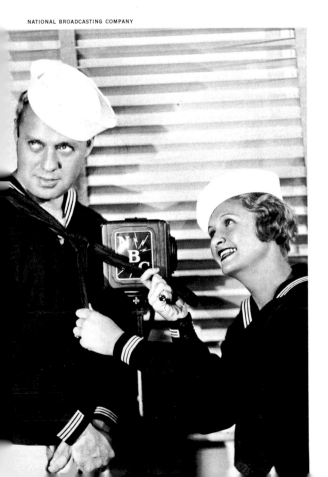

President Roosevelt speaks to miners in 1943

Jack Benny, not yet ''39,'' with Ethel Shutta in 1932

SMITHSONIAN INSTITUTION

systematic growth and development, his system itself was not flourishing. What he had not yet found was a completely reliable detector to replace the Branley coherer.

That search was preoccupying not only de Forest but also Marconi and a host of would-be inventors in England, Germany, France, and Russia. Every conceivable principle and effect was tested, whether understood or not.

The detector that survived can be traced to the day in 1883 when Edison put a metal-wire electrode with a positive charge into his light bulb. He discovered that electricity flowed from the glowing filament to the new electrode—across the space between them.

Twenty years later, John Ambrose Fleming, a British scientist serving as a consultant to Marconi, discovered that the "Edison effect" bulb could detect Hertzian waves in a completely new way. It was the first radio vacuum tube, the diode.

De Forest began experimenting, at that time, with a simple detector that contained a gas flame instead of an electric filament. In addition, he used a telephone receiver—with a battery of its own—and clearly heard the dot-dash wireless signals from a distant transmitter.

With laborious trial and error he added these two items to Fleming's diode. Later he tried a zigzag grid of wire between the filament and a metal-plate electrode to carry the incoming signal. And so de Forest made the first triode, patented January 15, 1907. He called it the "Audion" and—in retrospect—"the granddaddy of all the vast progeny of electronic tubes that have come into existence since."

But the progeny were not immediately forth-coming—partly because no one quite understood how the audion really worked. Some gases remained in the tube's partial vacuum. To de Forest, it seemed that the current could only flow from filament to plate through a transporting medium—the ionized gases. It was an obvious explanation—and incorrect.

By 1912 it had become clear to a number of investigators, including the distinguished researcher Irving Langmuir of General Electric, that the gases were unnecessary. The filament emitted electrons—particles with a negative charge—and the plate with its positive charge attracted them. Now it was evident that high-vacuum tubes would be far more efficient than the "soft" gassy tubes used by de Forest.

These new tubes permitted new experiments. De Forest and others discovered that with simple changes in circuitry the audion could amplify feeble electrical currents, thus immensely improving long-distance telephone transmission. But the really earth-shrinking properties of the audion were not revealed until August 1912, when de Forest found that with more complicated changes in the circuits he could obtain a range of musical tones in the receiver. He had added a "feedback" interconnection between the input and output circuits and created what radio engineers call "regeneration."

It was this single, completely unexpected result of an afternoon's tinkering that created the broadcast industry.

What de Forest had discovered was a very efficient method of producing continuous

De Forest taps code at his wireless exhibit, part of the 1904 World Exposition in St. Louis. He won the Grand Prize and Gold Medal—the fair's highest honors—for sending radio-wave messages to Chicago.

electromagnetic waves that could be modulated to carry the entire range of audible sound. The search for such a transmitter had been intense—for so long as transmission depended on spark gaps, wireless communication was limited to the dot-dash of the Morse code.

To parallel the leap from Morse to Alexander Graham Bell, to move from wireless telegraphy to wireless telephony (radio, to Americans), required continuous-wave transmission.

It was de Forest's audion that carried wireless from Marconi's faint *dit-dat* to Arturo Toscanini's "Symphony of the Air." It also provided the basis for one of the most protracted patent litigations in United States history.

In the stampede of invention that marked the early years of wireless, the regenerative circuit was discovered independently, within less than two years, by an Austrian, a German, and two Americans. One of the latter was Edwin H. Armstrong, a quiet-spoken man with a lumberjack's physique, who later developed FM. It was he who gave de Forest a tussle.

The battle began when it was declared that several patent applications, including de Forest's and Armstrong's, were in interference with each other. The first round went to Armstrong in 1923, the second to de Forest on appeal. And so the battlers swung away for ten long years, until finally the case reached the Supreme Court.

Although most foreign governments award patent rights to the first applicant, the United States seeks to award them to the first inventor, whether or not he outlegged the competition to the Patent Office. It thereby imposes on the judicial system the awesome requirement that individuals trained in the law make final judgments in some highly arcane areas of science and technology.

Such was the issue in *Radio Corporation of America* v. *Radio Engineering Laboratories, Inc.* When all the circuits had been laboriously traced and understood, the Justices returned to crucial testimony for their decision.

Armstrong had sworn that he had made his apparatus "about the first of October 1912." De Forest had sworn that his notebook's first description of the "so-called feedback circuit" was dated August 29, 1912.

On May 21, 1934, the Supreme Court found for the de Forest patents and nearly two decades of work, pleading, bitterness, and hope came to a close. In 14 years of litigation, the claims of de Forest had been upheld by seven courts, those of Armstrong by six.

All those years of running—and it took the Supreme Court to say he had won. He had seen his inventions bring immense fortunes to others, but not to himself. The professional glory he had striven for came to him dimmed by endless disputes over priority and business deals. An almost hopeless romantic, he had suffered two broken marriages.

And yet his unflagging energy and unquestioning enthusiasm propelled him into new pioneering challenges: the development of talking pictures, television, and finally, shortwave therapeutic devices. But each time he took his own course—and found himself alone.

In his lifelong drive to outrace the competition, had he outraced life itself?

Not quite. At the age of 57, de Forest met a young movie actress, Marie Mosquini, and after six weeks' wooing entered into marriage a third time. He had found the long-awaited companion who shared his deep love of music, poetry, and the beauty of the outdoors. They took long camping trips in state and national parks. De Forest spent less time in the laboratory and more in front of a massive high-fidelity phonograph he constructed for sheer enjoyment.

Radio gave him less pleasure; he detested its average programs and "crass" commercials.

But when he wanted to regain the heights, there was always Mount Whitney. He celebrated his 70th birthday by climbing it for the fifth time. In 1961, at the age of 87, he concluded the robust twilight of his life.

VICTOR R. BOSWELL, JR., N.G.S. STAFF

*Tiny dots of light—some 250,000—create this
Apollo 11 television picture as an electron beam
traces 525 horizontal lines on the tube 30 times a
second, striking blue, red, and green phosphorous
dots on each line. Atop the set, a 1934 "iconoscope"
camera tube rests beside its inventor V. K. Zworykin.*

VLADIMIR KOSMA ZWORYKIN
By Joseph J. Binns

"A FEW WEEKS after my enrollment in the Imperial Institute of Technology in St. Petersburg, the police arrived to put a stop to student rioting and disorders. We students barricaded ourselves in one of the campus buildings and were besieged for three days. Thus, at the very start of my college career, I had a confrontation with the police which, at that time in Russia, was quite common for a student."

The raconteur is Vladimir Kosma Zworykin, father of television. Fellow members of the National Academy of Engineering in Washington, D. C., are gathered to honor this engineer, scientist, and former student rebel with their coveted Founders Medal on April 24, 1968.

Dr. Zworykin went on to describe his repeated oscillations across the tenuous boundary between science and engineering, a pattern established early in life.

He was born June 30, 1889, in Murom, Russia, 200 miles east of Moscow, the youngest of seven children. At age nine he started spending his summers as an apprentice aboard the boats his father operated on the Oka River. He eagerly helped repair electrical equipment, and it soon became apparent that he was more interested in electricity than anything nautical.

Upon graduating from high school, he left for the University of St. Petersburg. After attending his first lectures, he resolved to become a physicist—a decision which lasted until his family heard about it. Their view was that in Russia's rising new industries, engineering offered a richer future than physics. Accordingly, Vladimir's father made him transfer to the Imperial Institute of Technology.

He loved the life of a student, even in the setting of restlessness and repression that characterized the decline of Czarist government. His courses included laboratory work in his great love, physics; faculty and students heatedly discussed the new atomic nucleus theory of Ernest Rutherford, the mysterious X-rays discovered in 1895 by Wilhelm Konrad Roentgen, Marie and Pierre Curie's current research in radioactivity.

Boris Rosing, a professor in charge of laboratory projects, became friendly with the young engineer and let him work on some of his private projects: Rosing was trying to transmit pictures by wire in his own physics laboratory. "Needless to say, I was soon there as his understudy," Dr. Zworykin recalls. "It was a glorious three years, and what a perfect school it turned out to be!"

Many scientists were attempting to extend man's sight, as the telegraph and telephone had extended his speech. An outstanding achievement was the mechanical scanner of the German inventor Paul Nipkow. Awarded the first television patent in 1884, the Nipkow disk had a series of square holes arranged near its edge.

In a Nipkow transmitter, one disk spun about 30 times a second between the scene to be televised and a phototube. Each hole admitted a bit of light from the scene to the phototube, which converted the brightness and darkness into variations in an electric current.

Sent by wire to a distant receiver and amplified, this current brightened or dimmed a light source. A second, identical, disk spun before the light source in exact synchronization with the one at the transmitter. An observer looking through the disk at the flickering light saw a series of bright-and-dark lines, which his eye translated into a continuous image.

From radio to radioactivity, all electromagnetic waves —including the slim rainbow of visible light (horizontal band)—belong to one natural family. A given wave's behavior depends on its frequency (number of times per second it oscillates) and its wavelength (distance between oscillation peaks). These

characteristics interrelate: the longer the wavelength, the lower the frequency; the higher the frequency, the shorter the wavelength. The figures on this scale, such as 10^6, constitute mathematical shorthand for very large or very small numbers. 10^6 means 1 followed by 6 zeros; 10^{-6}, 1 divided by 1,000,000.

This mechanical system was too insensitive to give practical results, although several decades later it found limited application—with improved disks, phototubes, amplifiers, and light sources —in the United States and Europe.

Rosing had been quick to see the advantages of an electronic system over a mechanical one. He and his young assistant experimented with a primitive cathode-ray tube, developed in Germany by Karl Ferdinand Braun.

"We made almost everything ourselves," Dr. Zworykin recalls, "becoming glassblowers to fashion photocells and amplifying tubes." And in 1910 Rosing exhibited a television system, using a mechanical scanner in the transmitter and the electronic Braun tube in the receiver. Although the system proved impractical because of limitations in the Braun tubes and the lack of suitable amplifiers, it fired Zworykin's imagination. (Boris Rosing did not live to finish his work. He was arrested during the Russian Revolution and died in exile.)

The lure of theoretical physics drew Zworykin to Paris after he graduated with honors and a scholarship in electrical engineering in 1912. There he studied X-rays under Paul Langevin. This work, plus the beginning of World War I, effectively kept him from pursuing his interest in television. As soon as war broke out, he made his way back to Russia. He was drafted immediately and found himself a few months later at Grodno fortress near the Polish border.

Grodno did not have a radio transmitter, although the components were on hand. When the communications officer discovered an electrical engineer among the recruits, he ordered Private Zworykin to build a transmitting station. The young squad leader commandeered a farmhouse and was sending messages to a nearby fort the very next day. When Zworykin reported that the orders had been carried out, the officer was at first incredulous and delighted, then suddenly angry. "It is against regulations," he shouted, "to send messages without encoding them."

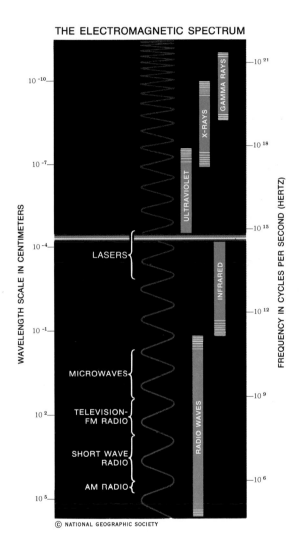

THE ELECTROMAGNETIC SPECTRUM

© NATIONAL GEOGRAPHIC SOCIETY

"You didn't give me the code," Zworykin pointed out.

"Certainly not," the officer replied. "It is against the regulations to give the code to a private." But when he cooled off they reached a compromise; Zworykin encoded and decoded every message in the officer's quarters, under his watchful eyes.

"Of course, I soon memorized most of the code and it wasn't always necessary to go to the officer's rooms to use the code book," says Dr.

Man, measure of all things in antiquity, and so portrayed with outstretched arms by Leonardo da Vinci, reaches from the closed world of the past into a limitless future. Visible light, a sliver of the electromagnetic spectrum, illumines his face. The remaining regions of radiation, inaccessible to his unaided senses, beckon him to exploration with the tools of his ingenuity. Television, its image borne on invisible waves from moon to Earth, shows Neil Armstrong

ARTHUR LIDOV

taking a ''giant leap for mankind''; a hand symbolically grasps the newly visited moon. Beyond appears the black disk of the sun in eclipse. A radiotelescope, amplifying microwaves, hears ''noise'' from a distant star, helping man to study such galaxies as the Andromeda Nebula. Orbiting 690 miles high, a Nimbus 4 uses the power of sunlight to relay infrared weather pictures to Earth. To its right, an Orbiting Solar Observatory pioneers in the study of the solar atmosphere. X-rays penetrate flesh to silhouette denser tissues. The large hand at upper right touches the nucleus of a living cell. Below it coil chromosomes containing the DNA molecule and its clues to heredity, probed by the smaller hand. Finally, at extreme lower right, appear viruses at the frontier of life. Using new ways to harness all ranges of radiation, man delves ever deeper into the atom's heart and farther through the alien reaches of space.

Zworykin, his eyes twinkling at the memory, "but it made him happy."

After a year and a half, he was transferred to Officers Radio School, won his commission, and began teaching electronics.

When the Russian Revolution began in 1917 and Zworykin saw that it would disrupt his scientific career, he decided to leave the country, but at first could not get permission. Then the United States refused him a visa. For months he wandered Russia to avoid arrest in the chaos of civil war between the Reds and the Whites.

When an Allied expedition landed in Archangel in September 1918 to aid Russia's northern defenses against the Germans, Zworykin made his way there. Pleading his case with an American official, the earnest young Russian told of the work he would do in developing television. He might have made sense to a physicist, but his recital seemed fantastic to the diplomat. Nonetheless, impressed by Zworykin's zeal and personality, he arranged for a visa.

Reaching London safely, Zworykin boarded a ship for America. Traveling first class as befitted an officer and gentleman, he discovered that the other passengers dressed formally for dinner. Not having a dinner jacket, he was chagrined when everyone stared at him each evening as he entered the dining salon. He could not know then this embarrassing situation might one day save his life!

Arriving in the United States in 1919, he soon joined the Westinghouse laboratory staff in Pittsburgh. After trying in vain to persuade his superiors to let him develop an all-electronic television system, he left the company. Eighteen months later he returned, and was given greater freedom to work on his many ideas, especially for television.

The chief components of a modern TV system are the camera at the sending point and the picture reproducer at the receiving point. The camera generates electrical signals that correspond to the brightness (and recently, color)

"Bonanza," starring Lorne Greene, reaches as many a:

in the transmitted scene. These signals are then processed and amplified so they can be sent over long distances to the receiver, where they are recovered and applied to the picture reproducer. This unit converts the signals into a visible image of the original scene.

In 1923 Zworykin applied for a basic patent on his electronic TV system and exhibited it to a group of Westinghouse executives. "By present standards the demonstration was scarcely impressive," he wrote in 1962. "The transmitted pattern was a cross projected on the target of the camera tube; a similar cross appeared, with low contrasts and rather poor definition, on the screen of the cathode-ray tube."

This performance indicated that his devices were sound — and needed, as he said, "tremendous improvement" to be useful. After the demonstration it was suggested that he devote time

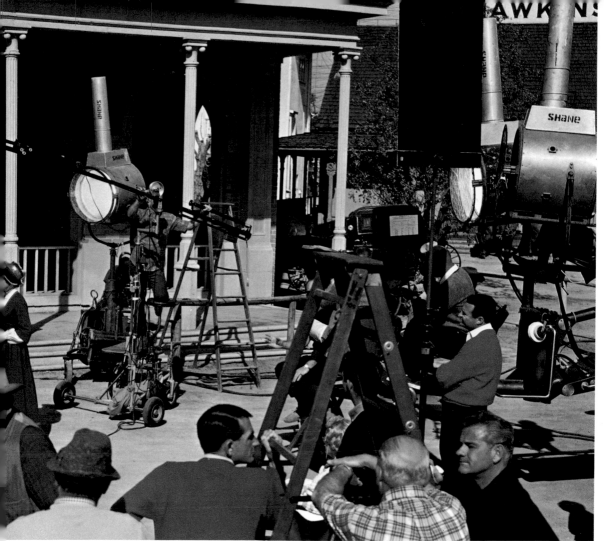

NATIONAL BROADCASTING COMPANY

million homes through Bell's microphone, Edison's camera, de Forest's triode, and Zworykin's picture tube.

to "more useful projects." He complied, but he continued to work on electronic television.

On November 18, 1929, at a convention of radio engineers, Zworykin demonstrated a television receiver containing his "kinescope," a cathode-ray tube with the principal features of all modern picture tubes.

And months before that historic session, another event had vitally affected the future of television. Dr. Zworykin met David R. Sarnoff, vice-president and general manager of the Radio Corporation of America. "Sarnoff quickly grasped the potentialities of my proposals," Zworykin later wrote, "and gave me every encouragement from then on to realize my ideas."

At the end of the year, a reorganization led to Dr. Zworykin's transfer to RCA in Camden, New Jersey. As the director of their Electronic Research Laboratory, he was able to concentrate

on making critical improvements to his system.

Zworykin and his staff began with the TV camera, producing in 1931 an improved version of his "iconoscope," a camera tube. A sensitive screen in its heart builds up an electric charge in proportion to the brightness of the scene at each point, while the scanning beam passes over the rest of the screen. In effect, the screen stores light for each point of the image and intensifies it more than 100,000 times.

Therefore, as the scanning beam hits each spot, it releases a much stronger picture signal than the old mechanical scanners provided.

The iconoscope completed the fundamental elements of a practical, all-electronic television system, and became the mainstay of TV broadcasting until after World War II. Zworykin's "storage principle" is the basis of modern TV.

Starting in 1934, ten years after he became a

LP record inventor Peter Goldmark demonstrates his video recording device that makes home TV libraries possible. Reel cassettes in his Connecticut laboratory hold color programs for replay on adapter-equipped standard TV sets. His camera photographs in color on black-and-white film and the adapter electronically restores the original hues on the TV screen. Other systems record on magnetic or laser-patterned tape.

naturalized U. S. citizen, Dr. Zworykin began to travel extensively for RCA. The beginning of World War II, in 1939, found him in Lebanon, where he managed to get on a plane bound for England. About to sail for the United States aboard the S.S. *Athenia,* he suddenly recalled that his tuxedo was in a trunk he had abandoned in Beirut. The memory of his earlier shipboard embarrassment overwhelmed him, and he decided to wait for a later vessel so he could buy formal wear and other clothing.

The *Athenia* never reached port. A German U-boat torpedoed and sank her off the Irish coast; among the missing were 28 Americans.

Back in the United States, Dr. Zworykin continued to work. In 1944 the German High Command named him as the inventor of a secret infrared device for the U. S. which could pierce fog and clouds, giving navigators "virtual maps by which to fly." Questioned by reporters, Dr. Zworykin laughed off the story.

Curiously enough, he was in fact working on infrared image tubes which would convert invisible infrared rays into visible light to allow humans to see in the dark without revealing their position. The device made possible the "sniperscope" and the "snooperscope," used by thousands of Allied soldiers in World War II.

After the war he continued to develop inventions and find new applications for electronics. He gave imaginative leadership to such projects as using TV pickup tubes and computers for weather forecasting; devices to direct planes on airport runways; computer-controlled traffic signals to prevent traffic jams; and experimental radio transmitters along freeways for automatic driving of specially-equipped automobiles, to eliminate highway accidents.

Upon his retirement in 1954, Dr. Zworykin could give more time to a long-standing interest as director of the Medical Electronics Center at the Rockefeller Institute in New York. At RCA he had directed a team that developed the electron microscope into a valuable research device. In-

stead of light focused by lenses, it uses a beam of electrons, focused by electromagnetic fields, to give a greatly magnified image on a fluorescent screen. Today it finds use in metallurgy and transistor-circuit production as well as medicine and microbiology.

A humanist, keenly aware that inventions do not automatically serve the public interest, he has suggested that television, the industry he helped to create, may one day strengthen the voice of the people. Questions of general concern might be presented to the Nation by radio and TV; answers would be transmitted by telephone to computers, with careful electronic safeguards against fraud. Such a polling system could provide, in Dr. Zworykin's words, "a method of assessing public opinion which is both rapid and accurate."

He is currently working on a system whereby information encoded on small cards can be sent over telephone lines. Dr. Zworykin envisions the "telecard" as a "medical passport"; it could carry a medical history, speed up record keeping and insurance payments, and furnish accurate, immediate data on epidemics.

He and his wife Katherine might well use the system to keep track of their five children and seventeen grandchildren, who pass in and out of the Zworykin home in Princeton, New Jersey, with erratic frequency.

The inventor finds the future of TV less unpredictable than the ramblings of his family. "Our first close view of the moon and the planets will, undoubtedly, be through the eyes of television," he wrote in 1954. Pictures sent from satellites in the mid-1960's proved his foresight.

Dr. Zworykin continues to imagine, create, and prophesy. His "brilliant mind has never waited for others: he has never stopped creating or innovating," said RCA President Robert W. Sarnoff in 1969. "Even in the fifteen years since his so-called 'retirement' from a remarkably productive career, he has accomplished more than many men do in a lifetime."

NATIONAL GEOGRAPHIC PHOTOGRAPHER JOSEPH J. SCHERSCHEL

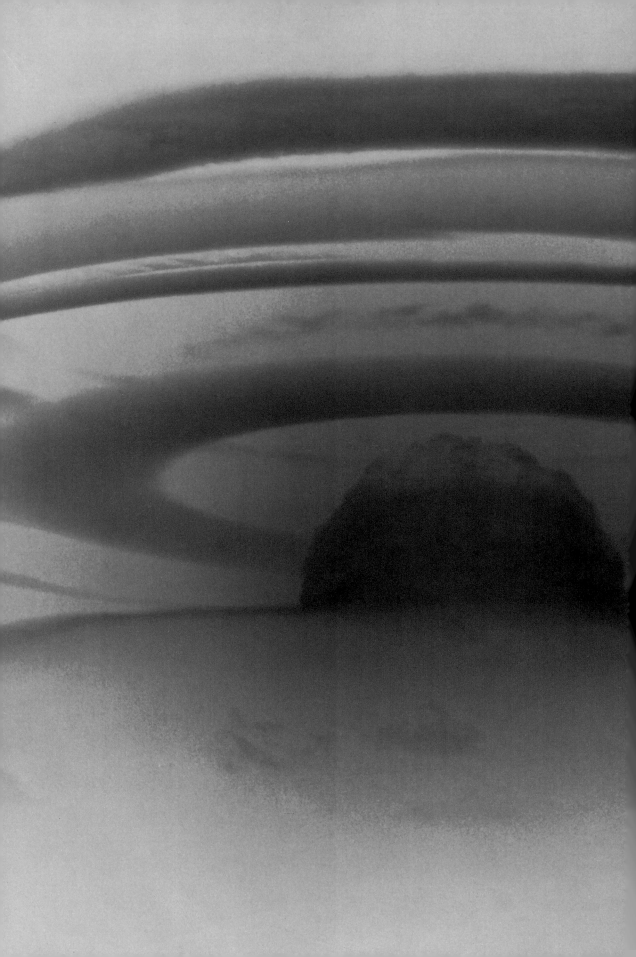

From the New World—to the New Universe

11

A S THE CENTURY TURNED, a German professor named Max Planck revised the accepted theory of the universe to account for a catastrophe that never happened—but should have according to the logic of classical physics. This logic demanded that radiant heat at any temperature above absolute zero have infinite energy. If that were indeed the case, a Franklin stove would turn into a minor sun. "Even the room you're sitting in," explains a helpful physicist, "would be a million times hotter than the interior of a star like Sirius."

No one knew what kept the world from disaster, but Planck had an explanation: The energy of radiant heat does not come in an unbroken flow but in discontinuous, incredibly tiny minimum units—a controlled situation. To his theory Planck gave the name "quantum" because he was dealing with distinct quantities of energy.

Rings of condensation, created by shock waves in moisture-laden air, spread for miles around a thermonuclear blast during tests in the Pacific—the result of matter becoming pure energy in an instant. The first such application of Albert Einstein's awesome equation E=mc² (energy equals mass times the square of light's velocity) occurred July 16, 1945, when a giant fireball lighted the dawn of the Atomic Age.

PHOTOGRAPHED ON EXTENDED-RANGE FILM, COURTESY CHARLES WYCKOFF, APPLIED
PHOTO SERVICES, INC., AND EG&G, INC., BEDFORD, MASS.

Fiery star trails swing around the North Star as the earth revolves during this two-hour exposure. Taming the whirling motion of infinitesimally smaller orbiting bodies, the cyclotron of Ernest O. Lawrence has let man explore the universe of the atom.

DAVID L. MOORE, N.G.S. STAFF

His concept was boldly extended by Albert Einstein, Niels Bohr, and others, to new theories of light and of atoms, to the new and eerie cosmos of quantum physics and a distinctive class of inventions to match such a setting.

In 1930, at the University of California at Berkeley, Ernest O. Lawrence started improvising a device to break up atoms for research. He began with a small pie-shaped gadget of copper, plate glass, and sealing wax. Discoveries made with his "cyclotron," and its descendants, now figure in nuclear weapons and power stations, in archeology (with carbon dating) and medicine.

Yet his countrymen were slow to associate Lawrence's work with that of men like Eli Whitney or the Wright brothers. As a writer for the *New York Times* observed: "... the servants of science invent as a matter of course. ... If Lawrence were what is called a *practical* inventor and his cyclotron were of any immediate commercial use, he would take his place beside

Watt, Arkwright, Bell, Edison and Marconi. ..."

Nobody could overlook another scientists' invention, the transistor, when it reached the market in the 1950's. From picnics in American parks to politics in the Middle East, from factories in Japan to spacecraft on the moon, transistors record by radio, television, and computer the triumph of the new solid-state physics — and of the three Americans who made it possible: John Bardeen, Walter H. Brattain, and William Shockley.

Of promise still unfathomed, Charles H. Townes's maser, an amplifier of invisible radiation, has opened a new realm called quantum electronics; it and its visible-light counterpart, the laser, already offer an amazingly wide range of functions — from making false teeth to finding evidence that synthesis of the chemicals necessary for the origins of life may have preceded the formation of planets.

If Americans look for a unifying theme in the story of their inventions, they can acknowledge first of all their debt in culture generally and technology specifically to Europe and the traditions of the Old World. For themselves, they might claim a zest for what is new, a readiness to scrap and rebuild and improvise and develop. America, said F. Scott Fitzgerald, "was a willingness of the heart."

Within half the lifetime of the United States, science and technology have, in effect, given man a new body. With fantastically sensitive instruments, he can reach far beyond the limits of his unaided senses: to detect the disturbance made by a few subatomic particles, to probe the mysteries of distant galaxies, to magnify the molecular structure of life, to record the order and timing of events taking place in a few trillionths of a second.

Within the lifetime of citizens not yet old, such works of genius have combined to enrich man's future — or, it may be, to wipe him off the face of a soiled earth. The risks and the solutions lie, as always, within his stubborn, inventive spirit.

Geniuses of 20th-century physics meet outside a laboratory at Berkeley, California, in 1938: From left, J. Robert Oppenheimer, theoretician; Enrico Fermi, 1938 Nobel Prize winner who combined theory and experiment; and Ernest O. Lawrence, who a year later would receive the Nobel Prize for his work on the cyclotron. His device accelerated charged particles of atoms around an atomic racetrack with precisely timed electrical impulses. Below, a cyclotron built in 1931 rests on Lawrence's notebook.

ERNEST ORLANDO LAWRENCE
By Michael Amrine

"A THOUSAND YEARS may pass before we can harness the atom, or tomorrow might see it with the reins in our hand." So it seemed to one scientist when the 1920's closed. As it happened, a great invention opened a new decade of discovery that culminated in the observation of nuclear fission by 1939. The device was one of classic simplicity and technical genius. In essence, it was the work and the life story of one American.

Far away from great cities, from famed laboratories and libraries, is a little South Dakota town called Canton. Out here the big-sky country begins; at night there are a million stars. In 1901 two boys were born here who were to change history by exploring the infinitesimal spaces within atoms. One was Ernest Orlando Lawrence, whose cyclotron has revealed as many new worlds in our time as the microscope or telescope in the past.

Charting the nature of the elements—and particularly, making artificial radioactive materials—Lawrence, his invention, and his associates deepened our understanding of the natural world. As salt is basic to the sea, as light and heat are to the stars over our earth, so are tiny radioactive throbbings a key to the nature of all matter. To a great extent it was Lawrence's machine that enabled man to take the pulse of these materials and to diagnose the behavior of atomic nuclei.

Through both parents, Lawrence came of Norwegian stock. His paternal grandfather was a teacher whose wife, an avid reader, never learned to write. Lawrence's mother was born in a sod house on the prairie in 1876. His father, Carl, rode into Canton on a bicycle to take a teaching job at the Augustana Academy offered him by Professor Anthony Tuve.

The Tuves lived across the street from the Lawrences and they had a boy, Merle, six weeks before Ernest was born. Close friends before they ever went to school, the two will be linked as long as men study atoms, for both became

LAWRENCE RADIATION LABORATORY, BERKELEY, CALIFORNIA

ARTHUR LIDOV

world-famous for their high-energy studies of subatomic particles and both produced new kinds of radiation.

Ernest, his younger brother John, and Merle grew up in a small-town America that Tom Sawyer might have recognized. Yet the relationship between man and nature was changing. The automobile, the airplane, the telephone were altering time and space, bringing under human control resources and energies Norwegian grandfathers could not have imagined.

Hooking telegraph keys to wireless sets, Ernest and Merle caught the new ideas crackling in the air. When they weren't experimenting with radio, they swam—or went skiing.

Lawrence's boyhood was flawed by a stammer, then touched by tragedy: the death of a 12-year-old cousin from leukemia. Ernest, John, and Merle all resolved to become doctors. (John, indeed, did, and with his brother developed ways to treat leukemia and other forms of cancer.)

Ernest intended to pursue medicine at the University of South Dakota, but an advisor there perceived a great scientist in the making. Once led into physics, he went on to the University of Minnesota, to join Merle Tuve; then to Chicago in 1923, to Yale in 1924. Like everyone else in physics, he was fascinated by the energies locked within the atom and preoccupied with learning more about atomic building blocks: protons, electrons, and other possible particles.

As a graduate student Lawrence was respected at Yale, had his work published in the highly prestigious *Physical Review,* and traveled through Europe to be welcomed at famous laboratories. At any age, the near-prodigy looked younger than his years: a National Research fellow at 24, an associate professor at the University of California at 27, a lithe figure with wavy light-brown hair and brilliant—some said cold—flashing blue eyes.

"Hard-working" and "possessed of unlimited

Physicists solemnly celebrate an event that changed the world—the first controlled, self-sustained nuclear reaction, December 2, 1942, at Chicago. After signing the wicker-covered Chianti bottle, Eugene Wigner passes his pen to Enrico Fermi. Leona Woods, only woman on Fermi's research team, waits to autograph the memento. Below, cadmium-coated control rods protrude from their "pile" of graphite and uranium. Power cables dangle in the foreground.

Inquisitive Ernest O. Lawrence, here tinkering with a cyclotron in 1934, often upset his staff with his enthusiastic operation of the delicate equipment, straining it almost to the breaking point. Below, a blackboard in the cyclotron control room at the University of California at Berkeley announces the celebration of his 1939 Nobel Prize in physics.

energy" are the first things said by many who knew him then. Some found him "the most normal of scientists," the only great physicist who could fit in at a Rotarians' convention.

Lawrence is one of the few giants of "the golden age of physics" who did not come from Europe—like Einstein, Enrico Fermi, Niels Bohr —or, at the least, from the East Coast. There older institutions looked down their noses at the laboratories slowly growing in some western universities. But when Lawrence settled at Berkeley in 1928, he led a tide of change in the geography of science. America was beginning to succeed Germany as the world capital of research; in time California would harbor more Nobel Prize winners than any other state, and more than most nations.

In later years, the pragmatic Lawrence was often contrasted with theoretician J. Robert Oppenheimer. Still later, the two held different views on national atomic policy; they were considered rivals, even political enemies. But once they were close friends. They worked on similar problems and were pursued by the same girls.

From his earliest days at Minnesota, Lawrence seemed fascinated with the thoroughly American idea of the biggest as the best, and with the twin problems of cracking the atom and achieving astronomically high energies. In none of this was he alone—other pioneering physicists were trying to fire electric "bullets" at the infinitesimal target: the atom.

As a young man Lawrence liked to work in his laboratory from 8 a.m. to midnight. He would play strenuous tennis, but partially as a duty— "got to keep in shape." He and a succession of associates begrudged the hours they had to spend eating or sleeping. For years he went with a girl who finally got tired of sitting around and waiting until some experiment was finished. Perhaps his way of relaxing had made matters worse—he spent many leisure evening hours browsing through scientific journals.

Later he met a Yale professor's daughter,

LAWRENCE RADIATION LABORATORY, BERKELEY, CALIFORNIA

Heavy-ion linear accelerator, right, "Hilac" for short,
shoots ions—charged particles—heavier than those
a cyclotron normally handles into target chemical
elements, changing their structure and creating
new elements. Electrical impulses between the tubes
boost the ions toward the target at the far end. Using
a cyclotron in 1940, physicists created neptunium
(in the microvial below), first man-made element.

LAWRENCE RADIATION LABORATORY, BERKELEY, CALIFORNIA

Mary (Molly) Blumer, whom he could not completely forget even for the new developments in
high-energy physics. They were married in 1932.

By 1929 Merle Tuve was producing radiation
that penetrated three inches of lead—and won
him a national prize. Many experimenters had
struck atoms with short bursts of energy. While
many American contemporaries were thinking
about making millions of dollars, Lawrence was
thinking about millions of volts.

One spring night, skimming a European
journal, he came on a minor paper by Rolf
Wideröe reporting experiments with ions—
atoms with positive charge due to the loss of
electrons. The Norwegian described hitting po-
tassium ions in vacuum tubes *twice* with the
same voltage. The result was that the ions moved
at *double the energy of the original voltage.*

Others had thought about giving atomic projectiles a series of pushes. Physicist Karl Comp-
ton likened the idea to pushing a child on a
swing: Give the child a small push at the proper
time and eventually the child will swing quite
high even though each small push would only
lift him a short distance. Thus, by the principle
of resonance, small but periodic stimuli produce
a large effect. The whole earth has since re-
sounded to the effects of that spring night.

Wideröe's paper fired Lawrence's scientific
intuition—he immediately conceived the principle of the cyclotron. That night in the library
he drew plans for boosting ions around a circular
course. A magnetic field or region of force could
bend their path and they could be accelerated
by resonance, by small energy boosts at precisely chosen instants.

An alternating electric field, changing at a
fixed rate, can provide such boosts of energy.
Lawrence saw that as the ions went faster, in
larger circles, they would pass the booster line
after the same interval because the higher
speeds compensate for the greater distances.
Theoretically, if the ions went around a million
times and were hit precisely right each time, a

one-volt push would have a million-volt effect. Traveling in a near-vacuum within the cyclotron, the ions would not lose appreciable energy. They would continue orbiting, much as the earth —carrying Canton, South Dakota—orbits through the vacuum of space around the sun.

Ernest had learned as a boy how the earth follows an orbit in space. His work and that of his contemporaries showed how particles could be energized into swift trajectories, then shot past the orbiting electrons into the hearts of atoms.

As a target, he said, an atom's nucleus is "like a fly in a cathedral"—and a busy fly at that. But the target could be hit if one could put millions of projectiles in a resonant accelerator.

That night, Lawrence told a colleague, "I'm going to bombard and break up atoms!" The next evening he told friends, "I'm going to be famous!" But even the confident or cocky Lawrence was probably not certain how neatly his instrument would work. Associates, like M. Stanley Livingston, refined the concept and contributed engineering expertise. It took almost a year for the first "magnetic resonance accelerator" to be proven. In laboratory slang the device became "cyclotron," the name that stuck.

Meanwhile, at a Carnegie Institution laboratory in Washington, D. C., Merle Tuve and his colleagues were working on a different method to obtain brief potentials as high as 5,000,000 volts. At Princeton, young Robert van de Graaff was thinking of a huge generator that would

High-energy protons—atomic particles—and their progeny trace patterns resembling abstract art. An accelerator produced these designs by sending particles crashing into a tank of liquid hydrogen, generating trails of bubbles—telltale fingerprints of atomic collisions. From the photographs scientists decipher reactions between particles, furthering man's knowledge of the titanic forces in atomic structures.

Diagram of the cyclotron principle (below, at left) shows two electrically charged half-cylinders, or "dees," their polarity switched repeatedly by an oscillator to attract and repel ions and spin them outward ever faster. A deflecting electrode directs particles to the target. In a section of a future accelerator (at right), bending magnets force the beam into a circular path. Focusing magnets keep ions tightly packed in a fusillade of atomic "bullets."

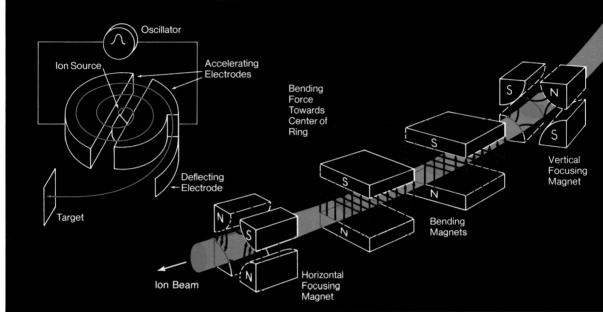

BEVERIDGE AND ASSOCIATES, INC.

ultimately produce 7,000,000 volts from static electricity.

Lawrence's devoted young team spent many frustrating months "tuning" their machine to achieve what they were certain of in theory. Glass plates in the first contraption shattered under testing. In 1931 Livingston, then a graduate student, demonstrated the predicted resonance in a brass model with a 4-inch round resonant chamber. He then built the "11-inch" that produced 1,200,000-volt protons in January 1932. When Lawrence came into the lab and heard the news he literally danced for joy.

In spite of the limited budgets of the Great Depression, Lawrence obtained funds for the bigger and stronger accelerators that followed the first crude versions: a "27½-inch" in 1934; the "37-inch" in 1937, performing well at 8,000,000 volts. Like the telescope, the cyclotron seemed

to generate knowledge. Physicists systematically bombarded one element after another, using one projectile after another.

In January 1934 Frédéric and Irène Joliot-Curie discovered artificial radioactivity in Paris, and the cyclotron provided efficient ways of inducing it. Then in just eight years scientists found 360 radioactive isotopes—substances produced by adding neutrons to the nuclei of natural elements. The majority—fully 223—were discovered with the cyclotron, 120 of them in Lawrence's laboratory.

John Lawrence, now a physician, joined his brother's staff at Berkeley in 1935 to lead medical research. He used radiosodium, radiophosphorous, and other isotopes to treat cancer.

Early in 1939 physicist Lise Meitner—who had escaped from Hitler's Europe to Sweden—announced the success of German experiments in

Slabs of concrete shield the Berkeley bevatron, descendant of Lawrence's cyclotrons. The massive accelerator whirls 800 billion hydrogen nuclei the equivalent of 300,000 miles in 1.8 seconds. Below, the blue beam of helium atom nuclei from a cyclotron bombards a quartz vessel of pure water, transforming oxygen in the water into radioactive fluorine, used to diagnose brain and bone tumors.

nuclear fission: the breaking apart of atoms. When her cable reached America, Merle Tuve and others confirmed that uranium atoms could be split. Tuve quite simply told reporters that this marked a turning point in human history.

That fall, Lawrence won the Nobel Prize, most coveted in science, for his cyclotron and for producing artificial radioactivity. Because of the war, he received the award at Berkeley.

In 1940 Lawrence turned to the Rockefeller Foundation for assistance in building a 184-inch cyclotron that required the world's largest magnet: 4,500 tons' worth. Receiving $1,150,000, Lawrence became a pioneer of "big science"—innovator of the megavolt and the megabuck.

In 1941 his laboratory—using a magnet from a cyclotron—produced the first appreciable amount of enriched uranium 235, the fissionable element in the first atomic bomb. Ironically, it was collected on December 7—the day the Japanese attacked Pearl Harbor and triggered the United States' entry into a world war that would end in 1945 when U-235 had brought holocaust to Hiroshima and Nagasaki.

During the war and after, Lawrence devoted himself to new roles as physicist and patriot. He brought desperate urgency to the Government's project for building a nuclear-fission bomb. In the production plant at Oak Ridge, Tennessee, his electromagnetic "calutron" process was used to extract fissionable isotopes from their stable cousins—although two other methods later outperformed it. Of physicists consulted on use of the first A-bomb, he fought longest for a demonstration that might convince the Japanese military without harming civilians.

In the cold-war years, when issues of nuclear strategy and national security dominated public concern, he became an influential—and controversial—science advisor to the U. S. Government. Behind the scenes he proposed the use of radioactive waste as a weapon in itself. He tried unsuccessfully—and spent millions of tax dollars in the process—to build a super accelerator

to turn tons of the Atomic Energy Commission's waste products into fissionable plutonium. Some thought it sinister; a fellow Nobel laureate said worse: "Silly."

To his critics it seemed that Oppenheimer, Einstein, and others voiced the conscience of science, while Lawrence remained preoccupied with super weapons and super scientific feats, as if these could guarantee man's progress.

Yet all his life he loved physics, loved his instruments, worked at all hours, labored with co-workers at midnight, called them at five in the morning—"Woke up thinking about last night . . . all wrong . . . what we need to do is this."

One colleague said they had to rig governors for their apparatus, because Lawrence always wanted to "test equipment to its breaking point." In the end he did that to himself.

Unexpectedly at 57, comparatively young, full of the honors but shadowed by the bitterness of the Atomic Age, Lawrence died in California from ulcerative colitis, an illness probably related to his driving nervous energy. In the cause of peace, he had insisted on going to Switzerland for a conference with Russian experts; there his endurance finally failed.

His life story—ending in a decade of fear and suspicion—raises questions not raised in times of reason. We are not sure if our whirling world will survive. If it does not, its destruction may result from man's misuse of knowledge generated by Lawrence, his cyclotron, his fellow physicists. If it does survive, spinning on through space, part of it will always be named for him: the element *lawrencium*.

Looking at this all-American boy, one feels that sense of awe and mystery which Einstein insisted should remain with even the most knowledgeable and logical of men. What had inspired Lawrence and his forebears and his fellow citizens with faith they could make tomorrow better? Man has not built the instrument that can penetrate the greatness and smallness which is a mystery within each of us.

VICTOR R. BOSWELL, JR., N.G.S. STAFF (ABOVE); LAWRENCE RADIATION LABORATORY, BERKELEY, CALIFORNIA

Inventors at play: John Bardeen, with grandchildren Charles and Karen, and Walter H. Brattain (bottom) share credit for an electronic revolution. The transistor they developed in 1947 with physicist William Shockley fundamentally changed the design of electrical tools—from hearing aids to communications equipment to giant computers—by replacing fragile vacuum tubes with much smaller, sturdier, solid parts that use little power and last for decades.

JOHN BARDEEN, WALTER BRATTAIN, WILLIAM SHOCKLEY
By Robert W. Holcomb

O N DECEMBER 23, 1947, a group of men gathered around a makeshift device in the Bell Telephone Laboratories near Newark, New Jersey. Before them lay a chip of silvery germanium about 1/16th-inch thick, with two fine wires sticking out of the top like the antennae of an insect. It was the first transistor, a Christmas gift to the world for which John Bardeen, Walter H. Brattain, and William Shockley were to win the Nobel Prize in 1956.

The men were sure the device would work and were excited at their success, but their mood was subdued. "I don't think anyone then could have foreseen everything that was to result from the discovery," says Dr. Bardeen, "although we knew it was tremendously important."

For years Bell Labs had been searching for a replacement for the hot, fragile vacuum tube. At last they had found it.

Men spoke into a microphone, and the inventors and other staff members listened to voices coming from a loudspeaker, amplified by the device on the table. With a graceful sense of history, one of the sentences repeated that day was, "Mr. Watson, come here. . . ."

Bardeen, Brattain, and Shockley had managed to make the electrons in a tiny sliver of solid matter perform the services previously rendered by the electrons in the vacuum tube— the progeny of de Forest's audion.

Like the vacuum tube, the transistor permits a very small current in one circuit to control a much larger current in another circuit. However, a million transistors could fit into one vacuum tube; they generate much less heat; nothing in them has to wear out; they require no warmup time; they are cheap to operate.

It was as if someone today would discover how to make a TV set that hangs flat on the wall like a picture, runs for a year on one flashlight battery, never wears out, and sells for $10.

Without transistors many jobs would be difficult, and some impossible. If a modern computer could be made with vacuum tubes, its

COURTESY JOHN BARDEEN

Wisconsin-born physicist opens birthday gifts.

BELL TELEPHONE LABORATORIES, MURRAY HILL, N.J.

Physicist and teacher on the golf course

cost would be a hundred times greater—and the tubes would generate so much heat it would take a river like the Niagara to cool them.

Unlike many earlier inventions which were essentially the work of one man, the transistor was invented by a team with a definite objective in mind. Bell Labs knew what they wanted—a way to switch and amplify signals without vacuum tubes or mechanical switches—and were certain they could find it. "The transistor came about," wrote Brattain in 1968, "because Bell Laboratories undertook to develop fundamental knowledge to a stage where human minds could understand phenomena that had been observed for a long time." The transistor is a striking example of the "corporate invention," itself largely an innovation of the 20th century.

Nonetheless, each individual on the team played a unique and crucial part in the invention. Accidents played a role, too.

For example, it was almost by accident—a temporary lack of space for staff—that John Bardeen joined the team. In 1945 he had completed a wartime assignment for the Navy and joined Bell. There he moved into an office already shared by Walter Brattain and Gerald Pearson. Given a free hand to work on whatever caught his fancy, he chose to join them."

"I got into physics rather indirectly," Bardeen recalls. At the age of 25, he had taken just enough courses in this science to whet his appetite. So in 1933, with both a bachelor's and a master's degree in electrical engineering from the University of Wisconsin, he took up the study of physics at the graduate level. In the depths of the Depression, he scraped together his savings from three years' employment and went to Princeton, where he received his Ph.D. in mathematical physics in 1936.

Born on May 23, 1908, near the campus of the University of Wisconsin, where his father was dean of the medical school, John Bardeen had been reared in an atmosphere that made the laboratory a natural place of work for him.

Walter Brattain had already been at Bell Labs for 15 years when Bardeen came to work there. Brattain was born on February 10, 1902, at Hsiamen (Amoy), China, where his father was teaching at a private school for Chinese boys. But his parents soon returned to Washington State, where they had grown up in pioneer families. His father took out a homestead near Tonasket and built up a cattle ranch.

"It was a normal, close-knit family," Dr. Brattain remembers. He loved outdoor life, and dropped out of high school for a year to work on the ranch. But he hated the farming part: "Following three horses and a harrow in the dust," he says, "was what made a physicist out of me."

He returned to school the following year.

He graduated from Whitman College, earned his master's degree at the University of Oregon in 1926, won his Ph.D. at the University of Minnesota three years later. Before joining Bell Labs in 1929, he worked in the radio section of the National Bureau of Standards—"the only non-teaching job I was offered; I was a very green Ph.D., and didn't feel qualified to teach."

The third member of the team, William Shockley, was born in London on February 13, 1910—his parents were living abroad. His grandfather was a New Bedford whaling captain, his father a mining engineer. During Nevada gold-rush days his mother held a Federal appointment as a deputy surveyor of mineral lands.

Shockley grew up and received his early education in California. While he was still in grade school, his interest in physics was aroused by a neighbor who taught at Stanford University. One of his early hobbies was magic; occasionally he would put on an amateur show for the neighborhood. (He still keeps a trick or two up his sleeve to lighten a lecture.) He earned his bachelor's degree from the California Institute of Technology in 1932 and his Ph.D. in physics from the Massachusetts Institute of Technology in 1936, and immediately joined Bell Labs.

This was the team that would succeed in

freeing electronics from the limitations of the vacuum tube. An eminent physicist who has known them for years says: "Bardeen is quiet, retiring in manner—a man of great firmness, though, and wisdom. Brattain's a peppery type, expresses himself freely; you can recognize the style of an ex-cowpoke. Shockley likes to pull surprises, to do the hard things—he enjoys rock-climbing. He likes to point out things he thinks others have missed, and he doesn't mind being at odds with others, in physics or not."

Together they began investigating semiconductors. These substances, notably germanium and silicon crystals, carry electricity better than insulators like ordinary glass but not so well as metals like copper. They had been in use before and during World War II—the crystal in a "crystal set" radio was a semiconductor; similar crystals detected radar signals.

Several theorists had already attempted to explain these strange substances, but nobody knew completely how or why they worked. Scientists reasoned that if semiconductors could do one job of the vacuum tube—detecting the information carried by radio or radar waves—they might also do another—amplifying. So Bell Labs formed several teams for research in solid-state physics in 1945, with Shockley directing the team that was to study semiconductors.

Brattain and Pearson, who had shared their office with Bardeen when he came to Bell Labs, had been investigating semiconductors since the 1930's. Neither Shockley nor Bardeen had much direct experience with semiconductors, but both had a good grasp of the quantum theory of solids, which predicts the actions of the parts of atoms. Others on the team included the physical chemist Robert Gibney and H. R. (Bert) Moore, a circuit-design expert.

In 1939 Shockley had had an idea for a solid-state amplifier using a semiconductor, but the war interrupted his project. Now it formed the basis for the team's early experiments.

Shockley's device consisted of a "slab" of

VICTOR R. BOSWELL, JR., N.G.S. STAFF, SMITHSONIAN INSTITUTION

Single dandelion seed covers a segment of a silicon wafer crowded with large-scale integrated circuits— electronic components and their connections formed directly on the wafer's surface. Each of the 16 chips shown here contains as many as 1,600 transistors. Invented in 1947, transistors modulate and amplify signals or act as switches. Before transistors, a 1946 calculator, ENIAC (with three of its 40 racks shown at left), contained some 18,000 vacuum tubes.

LOUIS ARBOLIDA, TRW INC.

semiconducting material, less than a thousandth of an inch thick, placed close to a metal plate but insulated from it. An electric field is associated with an electric charge, something like the magnetic field around a magnet. Shockley reasoned that this field, passing through the insulation, should affect the ability of the semiconductor to carry current. Thus, the signal pattern of a weak current in the metal, reflected in the field, would impress itself on a strong current flowing through the semiconductor. The strong current would repeat the pattern, thus amplifying the signal. The device was called a "field effect" amplifier because of the electric field controlling the strong current.

The trouble was, it didn't work.

For several months through the winter and early spring of 1946, Shockley would figure out new designs for a field-effect amplifier that the experimentalists would build and test. But in each case the effect was not there at all—or a thousand times smaller than expected. Other basic measurements proved just as puzzling.

"I believe the group as a whole slowly realized that these results were all of a piece," wrote Brattain several years ago, "and it was Bardeen who successfully explained them all."

"Your mind keeps on working on problems even when you're away from them for a while," says Brattain, "and sometimes the answer will just pop up." So he and Bardeen would occasionally retire to the golf course to relax, as the weeks passed with little progress.

Bardeen finally theorized that the atoms at the surface of the semiconductor were a key to the problem. Evidently the behavior of semiconductors was more complicated than anyone had thought; the former theories might be insufficient, or even dead wrong. Accordingly, the team stopped trying to make an amplifier and began to test Bardeen's theory. These experiments led to the invention of the transistor.

Brattain took up various investigations. He studied the electrical effect produced when light strikes a semiconductor surface. He also tried heating the crystal, and cooling it, measuring the voltage produced at different temperatures.

Then the weather interfered. In the sweltering, unairconditioned lab, during a New Jersey summer, water condensed on his apparatus as he cooled it, obscuring his observations. Reasoning that "if you can't beat them, join them," he immersed the apparatus in an insulating liquid. He noticed with interest that the liquid itself affected the results. He began a new series of experiments, immersing the apparatus both in insulating liquids and in conductors, such as solutions made with water. Though he did not realize it then, he was a step closer to the transistor.

Watching one of these experiments in November, Gibney suggested that Brattain run electricity through the conducting liquid into the semiconductor, to see if it affected the current produced by light. To their surprise, they found that by varying the electricity passing through the liquid they could not only decrease the current produced by light, but even change its direction. They had found Shockley's field effect!

Back on the trail of an amplifier, the team excitedly began trying to make one.

They built a model using a drop of liquid as one of the electrical contacts. Trying to replace the drop with a metal contact led directly to success. Testing such a model, they found it did amplify a little. But the currents flowed in directions opposite to those they had predicted.

Research physicists, however, learn to expect the unexpected. They found that a layer of insulation had been accidentally washed off. As a result, the device was simply a piece of semiconductor on a metal plate with two metal contacts close together on top of it.

Bardeen analyzed the results, and estimated that it would amplify much more if the contacts were about two thousandths of an inch apart. Brattain made the device by attaching a strip of gold leaf to a wedge of insulating material and cutting through the gold foil with a razor blade

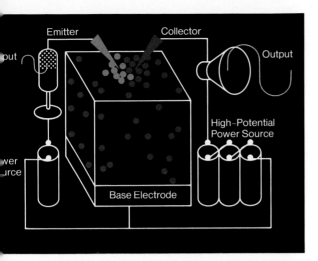

Emitter   Collector

put   Output

High-Potential
Power Source

wer
urce

Base Electrode

Early point-contact transistor amplifies by impressing patterns of a weak current onto a strong one. Sound —input—entering the microphone varies the weak current. Electrons—red dots—flow into the emitter, leaving "holes," or positive charge carriers—blue dots—in the transistor body. Attracted to the collector, these holes increase the flow of current from the high-potential power source. The loudspeaker converts the current variations into sound—output.

Silicon chips glow in the 1,922° F. heat of an electric oven. Precise amounts of impurities absorbed by the chips prepare them for use as bases for minuscule circuits containing hundreds of transistors and other electronic components. Below, young William Shockley, later one of the three inventors of the transistor, gallantly helps a playmate in distress.

BEVERIDGE AND ASSOCIATES, INC. (TOP); NATIONAL GEOGRAPHIC
PHOTOGRAPHER EMORY KRISTOF (CENTER); COURTESY WILLIAM SHOCKLEY

to make the two contacts. He then pressed the contacts onto a piece of germanium on a metal plate, and the first transistor was born.

Weak current flowing to the metal base from one of the point contacts—later called the emitter—changes a very small region near that contact from a poor conductor to a good one. As a result, the semiconductor allows a strong current to flow from the second electrode, the collector, to the base. Thus the pattern of variations in a weak current is impressed onto a strong one.

It was the complex flow of electricity within the semiconductor, and at its surface, that had baffled everyone for so long. To understand it required half a dozen fundamental new ideas, including conduction electrons and "holes" as carriers of positive charge. All had to be clarified before the transistor could become a reality.

For six months the invention remained a secret while patents were drawn up and improvements developed. The public announcement in July 1948 had little popular impact. The *New York Times* discussed it in 36 lines at the end of a "News of Radio" column on page 46.

Working out schemes to test Bardeen's ideas about surface states, Shockley thought of a second type of amplifier, the junction transistor. But exacting metallurgical problems had to be solved. Precisely one atom of impurity had to be added in the right place to each hundred million atoms of pure semiconductor. It was 1950 before Morgan Sparks and Gordon Teal at Bell Labs made the first practical models.

Almost all transistors now in use are junction devices, but recently a field-effect transistor— the type Shockley originally tried to develop— has come into use in modern integrated circuits.

Although transistor research remains very active, the three inventors who shared the Nobel Prize have gone on to other fields.

Brattain is proud to have been associated with Bell Labs, but he retained a preference for the Northwest all his life. Reaching mandatory retirement age in 1967, he promptly returned to

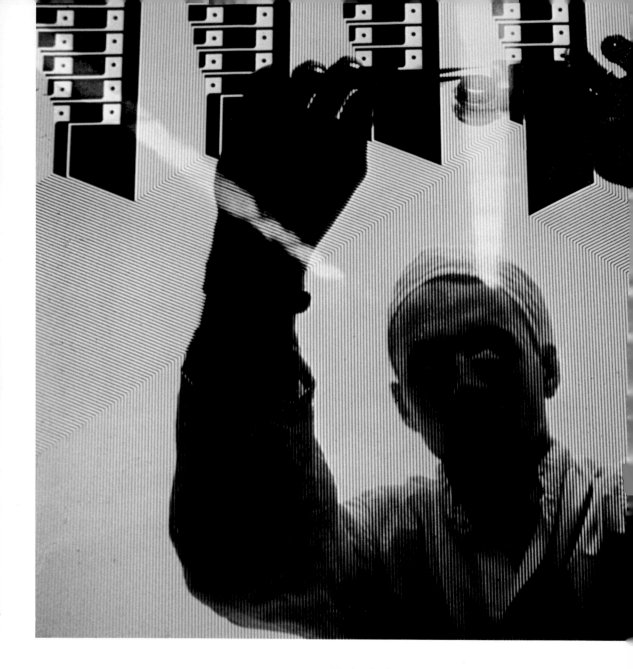

Big beginnings of tiny circuits: An IBM technician at Burlington, Vermont, inspects a computer memory circuit diagram on a huge sheet of film. Room-size cameras will take a series of photographs of it, shrinking it to a fraction of an inch. A similar process produces integrated circuits at TRW Inc., Redondo Beach, California, where a draftsman (opposite, upper) checks camera focus. At right, a computer-controlled plotter draws a circuit on film with a beam of light. Reduced 200 times, the diagram will control one of many metallic deposits forming thousands of electronic components on a silicon wafer about 1/7th-inch square.

ROBERT ISEAR

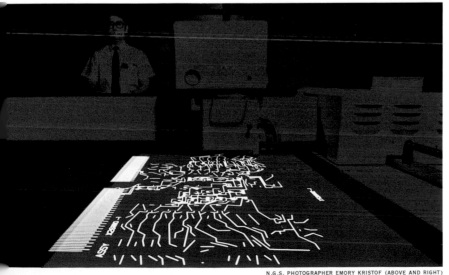

N.G.S. PHOTOGRAPHER EMORY KRISTOF (ABOVE AND RIGHT)

215

Lacelike "wiring" blankets a memory chip—as wide as seven fingerprint ridges—that does the work of 1,244 transistors plus another 1,200 miniature electronic components. Made by Fairchild Semiconductor in California, it will take its tiny but vital place in the brain of ILLIAC IV, a supercomputer being completed by the Burroughs Corporation. ILLIAC will provide a computer service for scientific research, eventually performing a billion operations a second.

Washington. He teaches at Whitman College, conducts biological research in the surfaces of cells, and takes time out regularly for golf.

Bardeen, the "physicists' physicist" of the three, finds his position as professor at the University of Illinois thoroughly congenial. He has made important contributions to pure physics, and maintains an interest in applied physics as a consultant and director of Xerox Corporation. Colleagues, who esteem him highly, elected him to his recent term as president of the American Physical Society. He and Brattain still meet and play golf occasionally, and Brattain describes him as "no experimentalist" at golf.

Shockley worked longest with transistors, establishing a semiconductor lab of his own after leaving Bell Labs in 1955. He is currently a professor of engineering science at Stanford University and executive consultant to Bell. On more than 80 patents, he appears as inventor or co-inventor of highly sophisticated devices.

The members of what Brattain calls "the best research team I was ever associated with" have gone their separate ways, but the balance that they brought to the invention of the transistor— a balance among basic physics, technology, and development—has followed it to the present.

Now a multibillion-dollar industry produces radiation-resistant transistors that control nuclear-power generators, and transistorized pacemakers that prolong the lives of heart patients. Transistors make it possible to reduce the size of other circuit components, and the microscopic integrated circuits that result, with thousands of transistors in each square inch, do thousands of jobs from processing a water bill to landing men on the moon.

Transistor radios have become so cheap that, says Brattain, "even the loneliest nomad on the steppes of Asia can have the news of the world just by twisting a dial. He doesn't have to learn to read. And once the common man has a fair chance to learn what's really going on, he has a chance to control his destiny."

NATIONAL GEOGRAPHIC PHOTOGRAPHER BRUCE DALE

Matching its creators' bright versatility, a krypton laser beam breaks on a diffraction grating before Townes (left) and Schawlow. Lasers produce man's most concentrated—and perhaps eventually most useful—light. Their beams can cut steel, carry telephone calls, and guide ships. At right, at the U. S. Army's Redstone Arsenal in Alabama, a 180-foot gas laser focuses a 2,500-watt beam already used to test nose cones and destroy swamp hyacinths.

Charles H. Townes, Arthur L. Schawlow

CHARLES HARD TOWNES
By Mary Ann Harrell

WITH A HINT of his teasing humor, Charles H. Townes likes to say that surprise is routine—for inventors and physicists, anyway. Some of the practical devices resulting from his own work prove his point:

A timepiece that gains only one second in 10,000 years. A radio-wave amplifier hundreds of times less noisy than any known, in preparation for yet-unborn satellite communications. A gauge accurate to one ten-thousandth the diameter of an atom. A new instrument for eye surgery. A new kind of drill. A new welding torch.

These devices are the now-famous lasers and their lesser-known forerunners, the masers. They all work by producing and controlling electromagnetic waves as these interact with molecules or with electrons in atoms; and they can be made to align sewer pipes, to take the temperature of a planet, to send telephone calls by pipes full of light, to cut polyester carpeting without raveling the edges, and eventually, perhaps, to power satellites in orbit.

They came into being not from trial and error, like the triode, not from a corporation's decision, like the transistor, but as by-products of Townes's research in pure physics.

They are contributions, by an indirect path, of an inventor who took a special interest in ammonia vapor (the pungent stuff in household cleansers) and had developed a taste for physics as an "indoor hobby" by way of a boyhood fascination with birds, insects, frogs, turtles, and fish.

Born in Greenville, South Carolina, July 28, 1915, Charles grew up exploring the fields, the swamps and creeks, the woods on farms his father owned. With his older brother Henry he shared a love of natural history; vacations frequently took them camping in the Blue Ridge Mountains, where bobcats still yowled at night.

He gives his mother credit for letting these interests flourish—"she put up with a lot of caterpillars in my room." His father brought home old clocks that the boys could take apart and reassemble. When school hours dragged, Charlie would divert his class with "magic"— whisking pens across the floor with "invisible" copper wire from the spark coil of a Model T.

Entering Furman University at 16, he took a full program of modern languages but planned a life in science. He might have gone into biology, he says, if Henry (now a distinguished entomologist) had not been so good at it. Physics fascinated him by its "beautiful logic" and its explanations of the world around him.

After graduating with both a B.A. and B.Sc. degree, he went on to Duke University, earning his master's and money to pay his way in the pinch of the Great Depression. At the California Institute of Technology, he lived on two meals a day and concentrated study; in 1939 he received his Ph.D. and a job offer from Bell Labs. He still remembers his astonishment that he would be so highly paid—$3,016 a year—to do physics.

Moving to New York City, he carried on research, took evening classes at the Juilliard School of Music, and changed apartments every

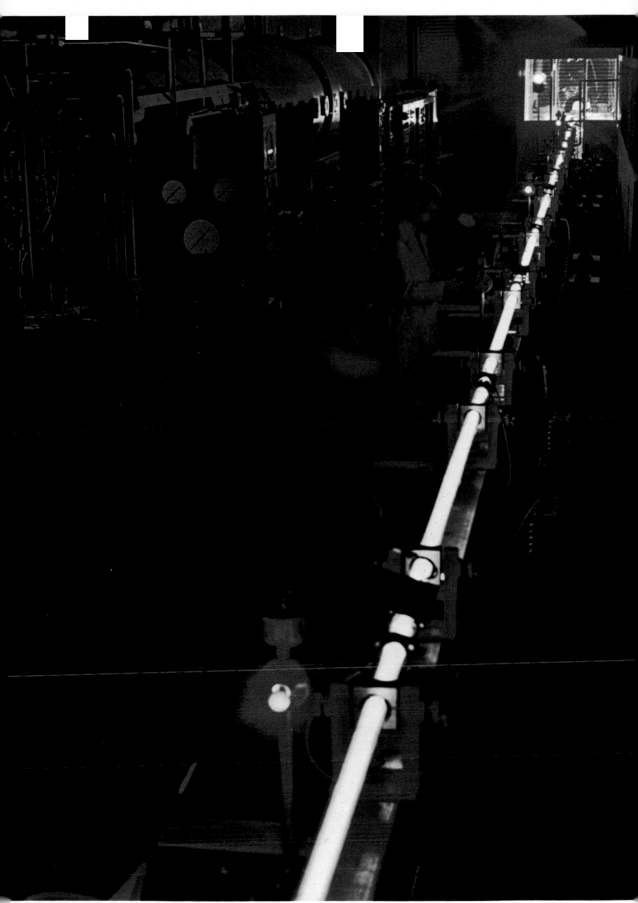

VICTOR R. BOSWELL, JR., N.G.S. STAFF (OPPOSITE); N.G.S. PHOTOGRAPHER EMORY KRISTOF

Townesian spectra: from Carolina summers to snowbound New England, from a scientific seminar to the sands of California. "Charlie" Townes sits on the arm of his grandmother's chair as a boy in Greenville, South Carolina. Decades later he drives a sleigh for two of his four daughters. In 1951 he tests a wet suit with Hugh Bradner, at right, who devised the garb to permit sustained diving in cold water. A man of wide curiosities and talents, Townes meets with American

three months to explore the city thoroughly. On a ski trip he met Miss Frances H. Brown from New Hampshire; they were married in May 1941.

A task forced on him by the crisis of the times —crash "hardware work" on radar—started him unknowingly toward his major inventions. The Army was using radar in the 3-centimeter band, radio waves about an inch long. It wanted radar gear that would weigh less and would function at a shorter wavelength, 1¼ cm. At that frequency, Townes warned, water vapor in the air would absorb most of the radar beam. "Older men overruled me," he says mildly. The system, hastily built, failed just as he had expected.

This led him to study the ways that water vapor and ammonia molecules absorb radiant energy. He designed and built new, unique types of spectroscopes for this research.

A spectroscope takes the fingerprints, so to speak, of atoms or molecules. Each atom or molecule absorbs or emits radiation at specific frequencies that form characteristic patterns called spectra. Townes used surplus gear from the ill-fated radar project for his instruments.

One of several men who opened the field of microwave spectroscopy just after the war, Townes was a leading authority when he joined the Columbia University faculty in 1948. He collected and labeled spectra as carefully as he had collected insects as a boy.

Meanwhile, engineers were eager to extend the use of radio to higher frequencies as short-wave channels grew more crowded. But they found the production of microwaves—much shorter than 1¼ cm—almost unmanageable. They needed good resonators, small metal cavities in which these weak waves could be amplified as they bounced back and forth. For such short wavelengths, the resonators would have to shrink to the size of a pinhead, too small to be built precisely, too small to function efficiently.

Einstein had published a clue to solving such a problem, in 1917, by using something provided by nature: the atom. He outlined a theory that

COURTESY MRS. HENRY K. TOWNES (ABOVE); COURTESY HUGH BRADNER

and Russian physicists during the First National
Conference on Quantum Electronics in Blooming-
burg, New York, in 1959. From left: James P. Gordon,
Nikolai G. Basov, Herbert J. Zeiger, Aleksandr M.
Prokhorov, and Townes. During his career Townes
has wrestled with scientific problems for the Pentagon
in peace and in war, and acted as chairman of a
committee on manned space flight for the
National Aeronautics and Space Administration.

COURTESY CHARLES H. TOWNES (BELOW); IRENE GOODENOUGH

Visible light earns the name "incoherent" by its jumble of wavelengths, or colors. In contrast, laser light acquires purity and power from its uniform wavelengths. In the prototype laser shown in sequence at right, chromium atoms in a ruby rod (A), its ends polished and silvered as mirrors, absorb light (B) from a flashtube (not shown). The light's energy "excites" the atom (detail), and an electron jumps temporarily into a wider orbit. The increased energy level makes atoms unstable, and they spontaneously give off photons (units of light, represented by red arrows) as the electrons return to their normal orbits (C). Some of the photons escape; others bounce between the mirrors and stimulate more atoms to emit light—the lasing process (D). As the light becomes more and more intense, within thousandths of a second, a pulse of coherent light (E) escapes through the partially silvered mirror.

an atom, by its natural properties, using energy stored within it, could amplify a microwave. He described the process and called it stimulated emission of radiation.

Many physicists knew this in theory, too well to find it very interesting. Engineers, in general, didn't know it at all. And during the 1930's the two groups usually went by separate ways.

World War II threw them together on all sorts of projects. They gained a new respect for each other. After the war, independently, several men of similar experience saw the possibilities of using molecules as natural resonators for microwaves. To all, including Townes, the problems involved seemed beyond solution at first.

Charles Townes shies away from dramatic scenes, and likes to emphasize the "small and stumbling steps" of sustained research that lead society along. "What really is marvelous," he has written, "is that scientific knowledge is so fruitfully accumulative."

But his own day of discovery is becoming famous in the story of invention.

As chairman of a Navy-sponsored committee on extremely short microwaves, Townes had come to Washington for a meeting in 1951. Waking at dawn, he slipped out of his hotel and found a bench in Franklin Square, where he sat admiring the red and white azaleas in full bloom, and turning over the microwave problem in his mind.

Six years of work suddenly blossomed into insight: a way to make ammonia molecules amplify 1¼-cm microwaves by the process Einstein had outlined. "I'd thrown away similar ideas many times before," he recalls, "but I spotted the weakness in my previous arguments." Rapidly he scribbled equations "on the usual back of an envelope." ("I had the numbers in my head," he remarks.)

But translating the equations into equipment was another matter. Two volunteers offered to try to build such a device: Dr. H. J. Zeiger, who helped in early stages, and another graduate student, James P. Gordon. Townes remembers

with pleasure "how low-key it was." Two Nobel laureates urged him not to waste money on something destined not to work.

Jotted notes for a speech sum up the story. "2 yrs & still stumped ... won't work—harder work—be stubborn, emotionally involved."

Early in 1954 the contraption worked; Jim Gordon burst into a seminar to announce the moment. Elated, Townes and his students celebrated over coffee. They coined a name: maser, for Microwave Amplification by Stimulated Emission of Radiation.

The maser represented a fundamental breakthrough in technology, of the type the Navy had hoped for. The new amplifying principle was the beginning and the heart of a field now known as quantum electronics.

Thereafter the low-key mood changed; the spirited competition that still marks this field was under way. New types of masers soon appeared.

In less than a decade masers were serving as atomic "clocks." Townes and colleagues used two to check—and verify—Einstein's theory of special relativity, in a test that has been called the most precise physical experiment in history. As amplifiers, masers extended the range of radiotelescopes, and made practical a radar exploration of the planet Venus.

Some of nature's best-kept secrets seemed to lie nearer, but still inaccessible, at even shorter wavelengths in the infrared region. Here research has lagged because these waves are comparatively weak. Scientists wanted devices for effective exploration.

Pointing out the difficulties involved, Townes and his brother-in-law, Arthur L. Schawlow, proposed an indirect and bold step: to skip the infrared and go straight to visible light. Here wavelengths are measured in millimicrons (ten-millionths of a centimeter), but men of genius since Isaac Newton have been recording the properties of light. A maser for visible light seemed possible; work in this familiar region might give clues to the infrared. In 1958 Townes

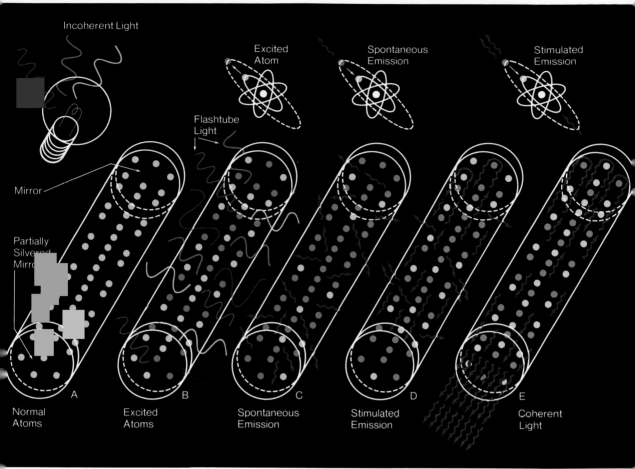

Incoherent Light

Excited Atom

Spontaneous Emission

Stimulated Emission

Flashtube Light

Mirror

Partially Silvered Mirror

A — Normal Atoms

B — Excited Atoms

C — Spontaneous Emission

D — Stimulated Emission

E — Coherent Light

BEVERIDGE AND ASSOCIATES, INC.

and Schawlow published the basic theoretical paper and suggested a variety of systems.

Several teams promptly began a sprint for a working model, and Townes faced what he soberly calls "a very difficult decision." Should he stay with fascinating research that already was arousing great excitement? Or should he accept a position to advise the Federal Government on scientific questions?

"I had quite a debate with myself," he says, "but ended up feeling it more important to tackle tough governmental problems, since working models could already definitely be foreseen." In 1959 he moved to Washington, taking up the role of expert consultant that—on various boards and committees—he sustains to this day.

In July 1960 the first burst of laser light flashed into the world. Townes and Schawlow had proposed it. It was produced by Dr. Theodore H. Maiman at Hughes Research Laboratories in Malibu, California, from a synthetic ruby. The name laser, for Light Amplification by Stimulated Emission of Radiation, promptly became

current—and a byword for power and precision.

Unlike the transistor, developed to meet specific industrial needs, the laser offered a new resource that challenged men to find uses for it. For a while it was called "an invention in search of applications," as improvements came pell-mell: a gas laser in 1961; giant-pulse lasers; the injection laser in 1962; tiny semiconductor lasers. By 1963, the year of the first liquid laser, some 500 research groups were busy with lasers in the United States alone.

Charles Townes received the Nobel Prize for physics in 1964, with Russian scientists N. G. Basov and A. M. Prokhorov, who had followed similar lines in fundamental contributions to quantum electronics. He holds many honors and awards. But he mentions—shyly—the thanks of friends whose sight has been restored by lasers used in retina surgery.

Schawlow, characteristically, tells a wry tale of his friend who had to make a second (successful) trip to the surgeon because the laser didn't function the first time. "Not working's

Oriental experiment uses lasers to project television pictures on a wall-size screen. In this Japanese system, three lasers—red, blue, and green—accept signals from color circuits of a standard TV receiver. Lasers aim their color beams at many-sided mirrors that rotate rapidly, synchronized with the receiver. Light from the mirrors bounces to the screen, enlarging the picture without losing brightness or clarity.

HITACHI, LTD.

World's highest-capacity memory unit stores a trillion bits of information, equal to a 20-line biography for each person alive. Made by Precision Instruments, Palo Alto, California, this $1,000,000 Digital Laser Recorder in a UNICON Computer "remembers" facts by piercing minute holes in data strips. Dr. Carl H. Becker tests the laser before it goes into operation recording findings for oil exploration.

N.G.S. PHOTOGRAPHER EMORY KRISTOF

the normal condition of lasers," he says, chuckling. "Like scientific equipment generally—not produced in sufficient volume to get thoroughly debugged. This has hampered laser development, but we'll get the kinks out with mass production."

As Schawlow noticed in 1961, laser beams concentrate so much energy on the surfaces they strike that they create strong electric fields. "With this," he says, "chemistry's a whole new ball game." He also foresees lasers, pleasantly silent, to replace rackety compressed-air drills.

To intensify a laser beam, special techniques concentrate it in time and focus it in place. The brand-new gasdynamic and chemical lasers promise the highest power yet, capable of trimming hot steel at plate mills in a few seconds per slice. Russian specialists have produced pulses of more than a trillion watts and with them forced neutrons out of the nuclei of atoms.

"Death-ray" weapons, however, seem mercifully remote: A very strong light beam disturbs the air it traverses, and these disturbances weaken it. Townes points out that if a strong beam *could* intercept ballistic missiles, its name ought to be "life ray." He also suggests that such beams could vaporize the growing clutter of space junk—dead satellites, old rocket stages—orbiting the earth. He grows enthusiastic about laser systems to dispel fog over airports.

Traveling often and widely to various meetings, he keeps a fast and disciplined pace, his mind on his work while he shaves or dresses or eats lunch. The effectiveness of a contemporary scientist, he says, depends vitally on the intensity of his interest, and this affects what he thinks of in such intervals.

Yet change obviously suits him.

"I've switched roles about every seven years," he observes. After seven years in administration,

Bright promise for man's future: Flashed across the Potomac River from the Washing-

including a stint as provost at the Massachusetts Institute of Technology, he moved to Berkeley to teach at the University of California and carry on research, now chiefly in astronomy.

With slight interest in "big-machine physics," he approaches the unknown with the intense patience of the naturalist observing living creatures. In this love for wildlife—as in his zest for languages, travel, music, and excellent food, and in his respect "for the intelligence of the American people"—he lives in the tradition of

Jefferson, whom his father admired and read.

Yet he suggests most of all that classic American figure, the wilderness scout—ranging far from the sedate cities, past the pioneer settlements, traveling fast, alert for the track of the wolf or the glint of a tanager's wing.

His wilderness is the universe.

Laser light may help to clarify baffling riddles in the study of subatomic particles, a disconcerting realm that even physicists call "weird." Maser amplifiers have increased more than one

ton Monument, a small laser's immutable glare bursts on a camera lens two miles away.

hundredfold the regions of the universe that radiotelescopes can explore.

And in fact, as Townes says, "the heavens are full of masers"—natural ones. From the Orion Nebula and many other directions, scientists have detected the unmistakable maser radiation of water vapor. Now other complex molecules have also been found in space: ammonia, formaldehyde, and hydrogen cyanide, which are associated with the early chemical stages of the evolution of life—a process that may have begun even before planets began orbiting a central sun.

Charles Townes expects that science and religion will draw closer to each other, science seeking order in the universe, religion seeking purpose and meaning. He cites the contributions of Ernest Lawrence and the Wright brothers as cases where "so-called hard-boiled realism wasn't real and dreams were." He likes to say that an indirect path is necessarily the best one in the most startling of discoveries and inventions—and in the pursuit of happiness.

Edison in his laboratory, about 1919

INDEX

Illustrations references, including legends, appear in *italics*.

Authors' Notes

MICHAEL AMRINE, science writer and editor, is author of *The Great Decision,* on the use of A-bombs against Japan. EDWIN A. BATTISON is Associate Curator of Mechanical and Civil Engineering at the Smithsonian Institution. JOSEPH J. BINNS has written on science, history, and music, is now managing editor of Robert B. Luce, Inc. ROBERT V. BRUCE, Ph.D., teaches at Boston University, is author of *Lincoln and the Tools of War,* is now completing a full-length biography of Bell. PAUL DOUGLAS, now teaching at Towson State College, has carried out research at the Smithsonian on Rumsey. Composer and critic of music, ROBERT EVETT has contributed to various magazines, is now a free-lance writer. After teaching English, JOHN GREENYA turned to free-lance work, is now president of Writing, Inc. ROBERT W. HOLCOMB has written for the American Institute of Physics, reported new research topics for *Science,* is currently free-lancing. HOWARD J. LEWIS is Director of the Office of Information of the National Academy of Sciences. CARROLL W. PURSELL, JR., Ph.D., teaches at the University of California at Santa Barbara, is author of *Early Stationary Steam Engines in America.*

ARTHUR P. MILLER, JR., is Assistant Chief of the Society's School Service. RONALD M. FISHER, MARY ANN HARRELL, and TEE LOFTIN SNELL are members of the Special Publications staff.

Additional references

The reader may wish to refer to the following NATIONAL GEOGRAPHIC articles for additional reading on American inventions and to check the *National Geographic Index* for other related material:

AGRICULTURE: Dorothea D. and Fred Everett, "Black Acres," November 1941; Frederick Simpich, "Farmers Keep Them Eating," April 1943; Jules B. Billard, "The Revolution in American Agriculture," February 1970.

AVIATION: Alexander Graham Bell, "The Tetrahedral Principle in Kite Structure," June 1903, and "Aerial Locomotion: With a Few Notes of Progress in the Construction of an Aerodrome," January 1907; Gilbert H. Grosvenor, "Dr. Bell's Man-Lifting Kite," January 1908; Alexander Graham Bell, "Prizes for the Inventor: Some of the Problems Awaiting Solution," February 1917; Rear Admiral Robert E. Peary, "Future of the Airplane," January 1918; "Air Conquest: From the Early Days of Giant Kites and Birdlike Gliders, the National Geographic Society Has Aided and Encouraged the Growth of Aviation," August 1927; F. Barrows Colton, "Our Air Age Speeds Ahead," February 1948; Carl R. Markwith, "Skyway Below the Clouds," July 1949; Frederick G. Vosburgh, "Flying in the 'Blowtorch' Era," September 1950; Vice Admiral Emory S. Land, "Aviation Looks Ahead on Its 50th Birthday," "Fifty Years of Flight," Hugh L. Dryden, "Fact Finding for Tomorrow's Planes," all December 1953; The Honourable Jean Lesage, "Alexander Graham Bell Museum: Tribute to Genius," August 1956; Kenneth F. Weaver, "Of Planes and Men," September 1965, and "Voyage to the Planets," August 1970; Harvey Arden, "World War I Aircraft Fly Again in Rhinebeck's Rickety Rendezvous," October 1970.

ELECTRICITY: Alexander Graham Bell, "Prehistoric Telephone Days," March 1922; F. Barrows Colton, "Miracle Men of the Telephone," March 1947; Albert W. Atwood, "The Fire of Heaven," November 1948; F. Barrows Colton, "Lightning in Action," June 1950; Lyman J. Briggs and F. Barrows Colton, "Uncle Sam's House of 1,000 Wonders," December 1951; Beverley M. Bowie, "The Past Is Present in Greenfield Village," July 1958.

ELECTRONICS: Robert Leslie Conly, "New Miracles of the Telephone Age," July 1954; Howe Findley, "Telephone a Star," May 1962; Thomas Meloy, "The Laser's Bright Magic," December 1966; Kenneth F. Weaver, "Crystals, Magical Servants of the Space Age," August 1968; Frederic C. Appel, "The Coming Revolution in Transportation," September 1969; Peter T. White, "Behold the Computer Revolution," and Ray Winfield Smith, "Computer Helps Scholars Re-create an Egyptian Temple," November 1970.

RUBBER: J. R. Hildebrand, "Our Most Versatile Vegetable Product," and Willard R. Culver and J. Baylor Roberts, "Rubber: From Trees to Tires and Toys," February 1940.

Composition for *Those Inventive Americans* by National Geographic's Phototypographic Division, John E. McConnell, Manager. Printed and bound by Fawcett Printing Corp., Rockville, Md. Color separations by Beck Engraving Co., Philadelphia, Pa.; Colorgraphics Inc., Beltsville, Md.; Graphic Color Plate, Inc., Stamford, Conn.; The Lanman Co., Alexandria, Va.; Lebanon Valley Offset Co., Inc., Cleona, Pa.; Progressive Color Corp., Rockville, Md.; and Stevenson Photocolor, Inc., Cincinnati, Ohio.

NATIONAL GEOGRAPHIC SOCIETY

WASHINGTON, D. C.

*Organized "for the increase and
diffusion of geographic knowledge"*

GILBERT HOVEY GROSVENOR
*Editor, 1899-1954; President, 1920-1954
Chairman of the Board, 1954-1966*

THE NATIONAL GEOGRAPHIC SOCIETY is chartered in Washington, D. C., in accordance with the laws of the United States, as a nonprofit scientific and educational organization for increasing and diffusing geographic knowledge and promoting research and exploration. Since 1890 the Society has supported 637 explorations and research projects, adding immeasurably to man's knowledge of earth, sea, and sky. It diffuses this knowledge through its monthly journal, NATIONAL GEOGRAPHIC; more than 27 million maps distributed each year; its books, globes, atlases, and filmstrips; 30 School Bulletins a year in color; information services to press, radio, and television; technical reports; exhibits from around the world in Explorers Hall; and a nationwide series of programs on television.

MELVIN M. PAYNE, President
ROBERT E. DOYLE, Vice President and Secretary
LEONARD CARMICHAEL, Vice President for Research and Exploration
GILBERT M. GROSVENOR, Vice President
THOMAS M. BEERS, Vice President and Associate Secretary
HILLEARY F. HOSKINSON, Treasurer
OWEN R. ANDERSON, WILLIAM T. BELL,
LEONARD J. GRANT, W. EDWARD ROSCHER,
C. VERNON SANDERS, Associate Secretaries

BOARD OF TRUSTEES

MELVILLE BELL GROSVENOR
Chairman of the Board and Editor-in-Chief

THOMAS W. McKNEW, Advisory Chairman of the Board

LEONARD CARMICHAEL, Former Secretary, Smithsonian Institution

LLOYD H. ELLIOTT, President, George Washington University

CRAWFORD H. GREENEWALT Chairman, Finance Committee, E. I. du Pont de Nemours & Company

GILBERT M. GROSVENOR, Editor, National Geographic

ARTHUR B. HANSON, General Counsel, National Geographic Society

CARYL P. HASKINS, President, Carnegie Institution of Washington

EMORY S. LAND, Vice Admiral, U. S. Navy (Ret.), Former President, Air Transport Association

CURTIS E. LeMAY, Former Chief of Staff, U. S. Air Force

H. RANDOLPH MADDOX Former Vice President, American Telephone & Telegraph Company

WM. McCHESNEY MARTIN, JR. Former Chairman, Board of Governors, Federal Reserve System

BENJAMIN M. McKELWAY Editorial Chairman, Washington Star

MELVIN M. PAYNE, President, National Geographic Society

LAURANCE S. ROCKEFELLER President, Rockefeller Brothers Fund

ROBERT C. SEAMANS, JR. Secretary of the Air Force

JUAN T. TRIPPE, Honorary Chairman of the Board, Pan American World Airways

FREDERICK G. VOSBURGH Former Editor, National Geographic

JAMES H. WAKELIN, JR., Former Assistant Secretary of the Navy

EARL WARREN, Former Chief Justice of the United States

JAMES E. WEBB, Former Administrator, National Aeronautics and Space Administration

ALEXANDER WETMORE Research Associate, Smithsonian Institution

LLOYD B. WILSON (Emeritus) Honorary Board Chairman, Chesapeake & Potomac Telephone Company

CONRAD L. WIRTH, Former Director, National Park Service

LOUIS B. WRIGHT, Former Director, Folger Shakespeare Library

COMMITTEE FOR RESEARCH AND EXPLORATION

LEONARD CARMICHAEL, Chairman
ALEXANDER WETMORE and MELVIN M. PAYNE, Vice Chairmen
GILBERT M. GROSVENOR, MELVILLE BELL GROSVENOR, CARYL P. HASKINS, EMORY S. LAND, THOMAS W. McKNEW, T. DALE STEWART, Senior Scientist, Office of Anthropology, Smithsonian Institution, MATTHEW W. STIRLING, Research Associate, Smithsonian Institution, JAMES H. WAKELIN, JR., FRANK C. WHITMORE, JR., Research Geologist, U. S. Geological Survey, CONRAD L. WIRTH, FREDERICK G. VOSBURGH, and PAUL A. ZAHL; BARRY C. BISHOP, Secretary on leave; EDWIN W. SNIDER, Secretary

Assistant Secretaries of the Society:
FRANK S. DELK, JOHN GOEDEL, JOSEPH B. HOGAN, RAYMOND T. McELLIGOTT, JR., EDWIN W. SNIDER

Leonard J. Grant, Editorial Assistant to the President; Edwin W. Snider, Richard E. Pearson, Administrative Assistants to the President; Judith N. Dixon, Administrative Assistant to the Chairman and Editor-in-Chief; Lenore W. Kessler, Administrative Assistant to the Advisory Chairman of the Board

SECRETARY'S STAFF: *Administrative:* Earl Corliss, Jr., Ward S. Phelps. *Accounting:* Jay H. Givans, George F. Fogle, Alfred J. Hayre, William G. McGhee, Martha Allen Baggett. *Statistics:* Everett C. Brown, Thomas M. Kent. *Payroll and Retirement:* Howard R. Hudson (Supervisor); Mary L. Whitmore, Dorothy L. Dameron (Assistants). *Procurement:* J. P. M. Johnston, Robert G. Corey, Sheila H. Immel, Margaret A. Shearer. *Membership Research:* Charles T. Kneeland. *Membership Fulfillment:* Geneva S. Robinson, Paul B. Tylor, Peter F. Woods. *Computer Center:* Lewis P. Lowe. *Promotion:* E. M. Pusey, Jr., Robert J. Warfel, Towne W. Windom. *Printing:* Joe M. Barlett, Frank S. Oliverio. *Production Control:* James P. Kelly. *Personnel:* James B. Mahon, Adrian L. Loftin, Jr., Glenn G. Pepperman, Nellie E. Sinclair. *Medical:* Thomas L. Hartman, M. D. *Translation:* Zbigniew Jan Lutyk

NATIONAL GEOGRAPHIC MAGAZINE

MELVILLE BELL GROSVENOR Editor-in-Chief and Board Chairman
MELVIN M. PAYNE President of the Society

GILBERT M. GROSVENOR Editor

FRANC SHOR, JOHN SCOFIELD Associate Editors

Senior Assistant Editors
Allan C. Fisher, Jr., Kenneth MacLeish, Robert L. Conly

Assistant Editors: Jules B. Billard, Andrew H. Brown, James Cerruti, W. E. Garrett, Edward J. Linehan, Carolyn Bennett Patterson, Howell Walker, Kenneth F. Weaver

Senior Editorial Staff: Lonnelle Aikman, Rowe Findley, William Graves, Jay Johnston, Stuart E. Jones, Robert P. Jordan, Joseph Judge, Nathaniel T. Kenney, Samuel W. Matthews, Bart McDowell; Senior Scientist: Paul A. Zahl

Foreign Editorial Staff: Luis Marden (Chief); Thomas J. Abercrombie, Howard La Fay, Volkmar Wentzel, Peter T. White

Editorial Staff: Harvey Arden, Thomas Y. Canby, Louis de la Haba, Mike W. Edwards, William S. Ellis, Alice J. Hall, Werner Janney, Jerry Kline, John L. McIntosh, Elizabeth A. Moize, Ethel A. Starbird, Gordon Young

Editorial Layout: Howard E. Paine (Chief); Charles C. Uhl, John M. Lavery

Geographic Art: William N. Palmstrom (Chief). *Artists:* Lisa Biganzoli, William H. Bond, John W. Lothers, Robert C. Magis, Robert W. Nicholson, Ned M. Seidler. *Cartographic Artists:* Victor J. Kelley, Snejinka Stefanoff. *Research:* Walter Q. Crowe (Supervisor); Virginia L. Baza, George W. Beatty, John D. Garst, Jean B. McConville, Dorothy A. Nicholson, Isaac Ortiz (Production). Marie L. Barnes (Administrative Assistant)

Editorial Research: Margaret G. Bledsoe (Chief); Ann K. Wendt (Associate Chief), Ledlie L. Dinsmore, Margaret L. Dugdale, Jan Holderness, Levenia Loder, Frances H. Parker

Geographic Research: George Crossette (Chief); Newton V. Blakeslee (Assistant Chief), Leon J. Canova, Bette Joan Goss, Lesley B. Lane, John A. Weeks

Phototypography: John E. McConnell (Chief); Lawrence F. Ludwig (Assistant Chief)

Library: Virginia Carter Hills (Librarian); Margery K. Barkdull (Assistant Librarian), Melba Barnes, Louise A. Robinson, Esther Ann Manion (Librarian Emeritus)

Editorial Administration: Joyce W. McKean, Assistant to the Editor; Harriet Carey, Virginia H. Finnegan, Winifred M. Myers, Shirley Neff, Betty T. Sanborne, Inez D. Wilkinson (Editorial Assistants); Dorothy M. Corson (Indexes); Rosalie K. Millerd, Lorine Wendling (Files); Evelyn Fox (Transportation); Carolyn F. Clewell (Correspondence); Jeanne S. Duiker (Archives)

ILLUSTRATIONS STAFF: *Illustrations Editor:* Herbert S. Wilburn, Jr. *Associate Illustrations Editor:* Thomas R. Smith. *Art Editor:* Andrew Poggenpohl. *Assistant Illustrations Editors:* Mary S. Griswold, O. Louis Mazzatenta. *Layout and Production:* H. Edward Kim (Chief). *Senior Picture Editors:* Charlene Murphy, Robert S. Patton. *Picture Editors:* David L. Arnold, William C. Latham, Michael E. Long, W. Allan Royce, Jon Schneeberger. *Research:* Paula C. Simmons. Barbara A. Shattuck (Asst.). *Librarian:* L. Fern Dame

Artists: Walter A. Weber (Naturalist), Peter V. Bianchi

Engraving and Printing: Dee J. Andella (Chief); Raymond B. Benzinger, John R. Metcalfe, William W. Smith, James R. Whitney

PHOTOGRAPHIC STAFF: *Director of Photography:* Robert E. Gilka. *Assistant Director:* Dean Conger. *Film Review:* Albert Moldvay (Chief); Guy W. Starling (Assistant Chief). *Photographic Equipment:* John E. Fletcher (Chief), Donald McBain. *Pictorial Research:* Walter Meayers Edwards (Chief). *Photographers:* James L. Amos, James P. Blair, Bruce Dale, Dick Durrance II, Otis Imboden, Emory Kristof, Bates Littlehales, George F. Mobley, Robert S. Oakes, Winfield Parks, Joseph J. Scherschel, Robert F. Sisson, James L. Stanfield. Lilian Davidson (Administration). *Photographic Laboratories:* Carl M. Shrader (Chief); Milton A. Ford (Associate Chief); Herbert Altemus, Jr., David H. Chisman, Claude E. Petrone, Donald E. Stemper

RELATED EDUCATIONAL SERVICES OF THE SOCIETY

Cartography — Maps, atlases, and globes: Chief Cartographer: Wellman Chamberlin; *Associate Chief:* William T. Peele. *Base Compilation:* Charles L. Stern (Supervisor), Charles F. Case, James W. Killion. *Name Compilation:* Donald A. Jaeger (Supervisor), Charles W. Gotthardt, Jr., Manuela G. Kogutowicz, David L. Moore. *Map Drawings:* Douglas A. Strobel (Supervisor), Robert W. Northrop, Tibor G. Toth, Thomas A. Wall. *Map Editing:* Ted Dachtera (Supervisor), Russel G. Fritz, Thomas A. Walsh. *Projections:* David W. Cook. *Layout and Design:* John F. Dorr, Harry C. Siddeley. *Revisions:* Richard J. Darley. *Archeology:* George E. Stuart. *Printing Control:* Richard K. Rogers. *Administrative Assistant:* Catherine M. Hart

Books: Merle Severy (Chief); Seymour L. Fishbein (Assistant Chief), Thomas B. Allen, Ross Bennett, Charles O. Hyman, Anne Dirkes Kobor, John J. Putman, David F. Robinson, Verla Lee Smith

Special Publications: Robert L. Breeden (Chief); Donald J. Crump (Assistant Chief), Josephine B. Bolt, David R. Bridge, Margery G. Dunn, Johanna G. Farren, Ronald M. Fisher, Mary Ann Harrell, Bryan Hodgson, Geraldine Linder, Philip B. Silcott, Joseph A. Taney

School Service: Ralph Gray (Chief and Editor of National Geographic School Bulletin); Arthur P. Miller, Jr. (Assistant Chief and Associate Editor of School Bulletin), Joseph B. Goodwin, Ellen Joan Hurst, Paul F. Moize, Charles H. Sloan, Janis Knudsen Wheat. *Educational Filmstrips:* David S. Boyer (Chief); Margaret McKelway Johnson, Bonnie S. Lawrence

News Service: Windsor P. Booth (Chief); Paul Sampson (Assistant Chief), Donald J. Frederick, William J. O'Neill, Robert C. Radcliffe; Isabel Clarke

Television: Robert C. Doyle (Chief); David Cooper, Carl W. Harmon, Jr., Sidney Platt, Patricia F. Northrop (Administrative Assistant)

Lectures: Joanne M. Hess (Chief); Robert G. Fleegal, Mary W. McKinney, Gerald L. Wiley

Explorers Hall: T. Keilor Bentley (Curator-Director)

EUROPEAN OFFICES: W. Edward Roscher (Associate Secretary and Director), Jennifer Moseley (Assistant), 4 Curzon Place, Mayfair, London, W1Y 8EN, England; Jacques Ostier, 6 rue des Petits-Pères, 75-Paris 2e, France

ADVERTISING: *Director:* William A. Boeger, Jr. *National Advertising Manager:* William Turgeon, 630 Fifth Ave., New York, N.Y. 10020. *Regional managers—Eastern:* George W. Kellner, New York. *Midwestern:* Robert R. Henn, Chicago. *Western:* Thomas Martz, San Francisco. *Los Angeles:* Jack Wallace. *Automotive:* John F. Grant, New York. *Travel:* Gerald A. Van Splinter, New York. *International Director:* James L. Till, New York. *European Director:* Richard V. Macy, 21 rue Jean-Mermoz, Paris 8e, France.

Flying saucers of 1910—sixteen cloth-covered disks attached to a spindly frame—never got off the ground.